COFINALLY COMPLETE METRIC SPACES AND RELATED FUNCTIONS

COFINALLY COMPLETE METRIC SPACES AND RELATED FUNCTIONS

Subiman Kundu
(retired) Indian Institute of Technology Delhi, India

Manisha Aggarwal
St. Stephen's College, University of Delhi, India

Lipsy Gupta
University of Missouri-Columbia, USA

World Scientific

NEW JERSEY • LONDON • SINGAPORE • BEIJING • SHANGHAI • HONG KONG • TAIPEI • CHENNAI • TOKYO

Published by

World Scientific Publishing Co. Pte. Ltd.
5 Toh Tuck Link, Singapore 596224
USA office: 27 Warren Street, Suite 401-402, Hackensack, NJ 07601
UK office: 57 Shelton Street, Covent Garden, London WC2H 9HE

Library of Congress Control Number: 2023934624

British Library Cataloguing-in-Publication Data
A catalogue record for this book is available from the British Library.

COFINALLY COMPLETE METRIC SPACES AND RELATED FUNCTIONS

ISBN 978-981-127-265-3 (hardcover)
ISBN 978-981-127-266-0 (ebook for institutions)
ISBN 978-981-127-267-7 (ebook for individuals)

For any available supplementary material, please visit
https://www.worldscientific.com/worldscibooks/10.1142/13316#t=suppl

Typeset by Stallion Press
Email: enquiries@stallionpress.com

Dedicated to

Professor Robert A. McCoy

Foreword

The main purpose of this monograph is to present a precise and detailed study of the (uniform) property called cofinal completeness in the frame of metric spaces. This property has been considered by many mathematicians since 1958 when it was first introduced by Corson. Cofinal completeness is one of the significant properties between compactness and completeness, which have been extensively investigated in the last two decades. In fact, the study of these intermediate properties constitutes nowadays an interesting and broad line of research in the setting of metric and uniform spaces.

This monograph contains in five chapters a careful review and an update of the most recent results obtained for cofinal completeness by many researchers including the authors themselves. In the first four chapters, this property is studied from different points of view. Namely, a wide number of characterizations are given for a metric space, as well as for its completion, to be cofinally complete. These characterizations are obtained in terms of different kinds of sequences, of special continuous and not continuous functions, of geometric functionals and also by using topologies in some of its hyperspaces. In the last chapter two stronger properties, also in between compactness and completeness, are considered. We are referring to the classical UC-ness defined by Atsuji and to the more recent notion of cofinal Bourbaki-completeness.

The present monograph is a very interesting work which will be useful for young and senior researchers in the vast topic of metric spaces as well as in related topics. It is well written and contains many examples in order to clarify the different concepts and characterizations appearing along the text. Also, it is worth highlighting the list of exercises after each chapter and the extensive bibliography.

Complutense University
Madrid (Spain)
November 2022

María Isabel Garrido

Introduction

The notion of a metric space was introduced by the French mathematician Maurice Fréchet in his doctoral thesis in 1906. However, the terminology was coined by the German mathematician Felix Hausdorff in 1914. Subsequently, the study of metric spaces was vigorously pursued by a host of Polish mathematicians in the 1920's. In metric spaces, we associate a real number to every pair of points. Broadly this real number signifies the distance between them. Thus, a metric on a set is an axiomatization of the fundamental properties of the distance function on the set of real numbers.

We all are familiar with one of the strong properties of topological spaces called compactness. When we have a metric structure, it is possible to gain many of the advantages of compactness with a weaker property. This gave rise to the concept of completeness of a metric space which was also introduced by Fréchet. Recall that a metric space is compact if and only if it is totally bounded and complete. Thus there is a huge gap between compact metric spaces and complete metric spaces. The role of these two classes of metric spaces in mathematical analysis is beyond question. Thus many researchers have been motivated to bridge the gap between them for a very long time. The classes of cofinally complete metric spaces, UC spaces, Bourbaki-complete metric spaces, cofinally Bourbaki-complete metric spaces are some better-known examples of the spaces that lies in between compact metric spaces and complete metric spaces.

A sequence (x_n) is Cauchy if for every $\varepsilon > 0$, there exists a residual set of indices $\mathbb{N}_\varepsilon$ such that each pair of terms whose indices come from $\mathbb{N}_\varepsilon$ are within ε of each other. If we replace "residual" by "cofinal" then we obtain sequences that are called cofinally Cauchy. Moreover, *a metric space is said to be cofinally complete if every cofinally Cauchy sequence in the space has a cluster point.* Certainly, the class of cofinally complete metric spaces lies in between the class of complete metric spaces and that of compact metric spaces. The primary objective of this monograph is to study such metric spaces extensively. Cofinal completeness was first considered implicitly by [Corson (1958)] and then by [Howes (1971)] in terms of nets and entourages. A few years later, [Rice (1977)] introduced the notion of uniform paracompactness for a Hausdorff uniform space X and subsequently in [Smith (1978)], the reviewer of Rice's paper for Mathematical Reviews, observed that uniform paracompactness is equivalent to net cofinal completeness for a Hausdorff uniform space. In the context of metric spaces, a metric space (X,d) is called uniformly paracompact if for each open cover $\mathscr{V}$ of X, there exists an open refinement $\mathscr{U}$ and $\delta > 0$ such that for each $x \in X$, $B(x,\delta)$ intersects only finitely many members of $\mathscr{U}$. In [Beer and Di Maio (2012)], it was shown that sequential cofinal completeness in metric spaces is equivalent to uniform paracompactness. In [Hohti (1981)], a nice equivalent characterization of a uniformly paracompact metric space is given in terms of uniform local compactness. Much later Beer cast a new light on cofinal completeness and gave various nice characterizations of cofinally complete metric spaces in [Beer (2008)].

Out of the four variants of complete metric spaces that we have mentioned before, the concept of UC spaces is the oldest one. It is well known that every continuous function from a compact metric space to an arbitrary metric space is uniformly continuous. But this property is not equivalent to compactness. It is a characteristic property of a larger class of metric spaces called UC spaces (also widely known as Atsuji Spaces). *A metric space is said to be UC if every real-valued continuous function on it is uniformly continuous.* The corresponding weaker form of Cauchy sequence used to characterize UC spaces is pseudo-Cauchy sequence. The notion of pseudo-Cauchy sequences was introduced in [Toader (1978)]. More precisely, a metric space is UC if and only if every pseudo-Cauchy sequence in the space with distinct terms has a cluster point, where a sequence is called pseudo-Cauchy if there exists a pair of terms arbitrarily close frequently. Since every cofinally Cauchy sequence is pseudo-Cauchy, every UC space is cofinally complete.

In 1958, the notion of finitely chainable metric spaces was introduced in [Atsuji (1958)] in order to characterize the metric spaces on which every real-valued uniformly continuous function is bounded. Atsuji proved: every real-valued uniformly continuous function on a metric space is bounded if and only if the metric space is finitely chainable. The class of finitely chainable metric spaces strictly lies in between the class of bounded metric spaces and that of totally bounded metric spaces (see also [Kundu *et al.* (2017)]). Finitely chainable metric spaces are also called Bourbaki bounded in the literature because these sets were considered in the book [Bourbaki (1966)]. To characterize finite chainability sequentially, Garrido and Meroño defined Bourbaki-Cauchy and cofinally Bourbaki-Cauchy sequences in [Garrido and Meroño (2014)]. Corresponding to these sequences, Garrido and Meroño introduced two new types of complete metric spaces, namely Bourbaki-complete metric spaces and cofinally Bourbaki-complete metric spaces. A metric space is said to be (cofinally) Bourbaki-complete if every (cofinally) Bourbaki-Cauchy sequence clusters in the space. The concept of cofinally Bourbaki-Cauchy sequences is similar to that of cofinally Cauchy sequences, in fact, every cofinally Bourbaki-complete metric space is cofinally complete. There is no direct relation between cofinally complete metric spaces and Bourbaki-complete metric spaces, but interestingly, the collection of cofinally Bourbaki-complete metric spaces is positioned in between the class of UC spaces and that of cofinally complete metric spaces.

In analysis on metric spaces, undoubtedly functions play a vital role. Two important classes of functions, namely the class of continuous functions and that of uniformly continuous functions are well-known to all of us. In [Snipes (1977)], Snipes studied the functions that lie strictly in between these two important classes of functions, which he called Cauchy-regular functions. A function between two metric spaces is said to be Cauchy-regular if it preserves Cauchy sequences, that is, it takes Cauchy sequences to Cauchy sequences. These functions were further investigated in [Borsík (1988, 2000); Snipes (1981)]. Since Cauchy-regular functions are continuous, they have also been widely known as Cauchy-continuous functions in the literature [Aggarwal and Kundu (2016); Beer and Garrido (2015, 2016)]. Recall that every continuous function on a metric space (X,d) is Cauchy-regular if and only if (X,d) is complete. Interestingly, every Cauchy-regular function on (X,d) is uniformly continuous if and only if the completion $(\widehat{X},d)$ is a UC space [Beer (1986)]. The role of Cauchy-regular functions is not only significant in the analysis of complete metric spaces but also in the spaces which are stronger than the complete ones. Analogously, some functions were defined which preserve the sequences acting as generalizations of Cauchy sequences, namely cofinally Cauchy, pseudo-Cauchy, Bourbaki-Cauchy and cofinally Bourbaki-Cauchy. These functions were useful in viewing the aforementioned stronger versions of complete metric spaces from a different perspective.

The significance of Lipschitz functions in mathematics has motivated many researchers to consider various Lipschitz-type functions [Beer *et al.* (2020); Beer and Garrido (2014, 2015, 2016, 2020); Garrido and Jaramillo (2008)], namely locally Lipschitz, Cauchy-Lipschitz, uniformly locally Lipschitz and Lipschitz in the small, in the past few years. The interesting fact is that the study of pairwise coincidence of these Lipschitz-type functions and the study of their stability under reciprocation (a never zero function with property P is said to be stable under reciprocation if its reciprocal also satisfies the property P) gives characterizations of the variants of complete metric spaces. Thus the study of cofinally metric spaces is inherently related to the theory of functions between metric spaces. Along with these functions, various properties of other functions such as Cauchy-regular, Cauchy-subregular, almost bounded, uniformly continuous and strongly uniformly continuous functions have also been investigated. Consequently, nice characterizations of cofinally complete metric spaces and the spaces which are stronger than cofinally complete spaces, namely UC spaces and cofinally Bourbaki-complete spaces, are obtained. The study of boundedness of various combinations of Lipschitz-type functions and other functions proved to be helpful in getting equivalent conditions for totally bounded and finitely chainable metric spaces. Characterizations of cofinally complete metric spaces in terms of some hyperspace topologies and function space topologies are also presented. The study of variants of complete metric spaces and all such functions is interesting in itself as well as it has nice connections with various other branches of mathematics such as convex analysis, optimization theory, fixed point theory, functional analysis and approximation theory.

The primary goal of this monograph is to give a comprehensive list of equivalent characterizations of cofinally complete metric spaces. Since cofinally complete metric spaces are complete, we also give a broad list of characterizations of the metric spaces whose completions are cofinally complete. Some equivalent characterizations of the metric spaces that are stronger than the cofinally complete ones, namely UC spaces and cofinally Bourbaki-complete spaces, are also presented. We study old results along with the recent developments on such spaces as much as possible. The discussion mainly concentrate on the following four kinds of characterizations of such metric spaces:

- (a) sequential characterizations,
- (b) characterizations based on certain functions,
- (c) characterizations in terms of geometric functionals, and
- (d) characterizations in terms of some hyperspace topologies and function space topologies.

The entire work of the monograph has been presented in five chapters.

In Chapter 1, we define the variants of complete metric spaces which we have mentioned earlier and various examples are given to illustrate the relation of these spaces with each other. Moreover, various Lipschitz-type functions, strongly uniformly continuous functions, Cauchy-subregular functions, functions preserving certain sequences, are briefly introduced. Some properties of Lipschitz-type functions and certain extension results related to continuous, Cauchy-regular, and uniformly continuous functions are mentioned as well.

In Chapter 2, cofinally complete metric spaces are studied in detail. To be precise, we present various characterizations of such spaces including characterizations based on sequences, functions and a geometric functional known as local compactness functional. Focus has been laid on CC-regular functions and their relations with other functions. In 2017, Keremedis defined almost bounded functions and AUC spaces [Keremedis (2017)]. Interestingly, almost bounded functions are nothing but CC-regular and the class of AUC spaces coincides with that of cofinally complete metric spaces. We study the stability of never zero continuous CC-regular functions under reciprocation and boundedness of various combinations of Lipschitz-type functions with CC-regular functions. In the entire process, nice characterizations of cofinally complete metric spaces are obtained.

In Chapter 3, we study metric spaces with cofinally complete completion; for brevity, it is said that such spaces have cofinal completion. Since cofinally complete metric spaces are complete, it is natural to pay attention towards the metric spaces which have cofinal completion. Such spaces are studied in terms of some properties like boundedness, uniform continuity of Cauchy-regular functions, Cauchy-Lipschitz function, and uniformly locally Lipschitz functions defined on them. Cauchy-subregular functions and their comparison with CC-regular functions have been studied. One of the key tools used here is the local total boundedness functional.

In Chapter 4, cofinally complete metric spaces are characterized in terms of some hyperspace and function space topologies. For a metric space (X,d), we consider the set $AC(X)$ of almost nowhere locally compact sets, which is a subset of the set of closed subsets in X. We consider some hyperspace topologies on the set $AC(X)$ which it inherits from the set of non-empty closed subsets in X as subspace topologies and characterize cofinally complete metric spaces in terms of relations to Hausdorff metric topology, proximal topology, Vietoris topology and locally finite topology on $AC(X)$. The equivalence of the topology of uniform convergence with the Hausdorff metric on some particular subsets of continuous functions and locally Lipschitz functions from (X,d) to (Y,ρ) is established, where the equivalence also characterizes the cofinal completeness of the space (X,d). Furthermore, we characterize the class of metric spaces for which the corresponding space $AC(X)$ equipped with the Hausdorff metric topology is cofinally complete.

In Chapter 5, we characterize UC spaces in terms of uniform continuity of various thin subclasses of continuous functions. Several characterizations of UC spaces and the spaces having UC completion are studied in terms of relations among various functions such as Lipschitz-type functions, continuous functions, CC-regular functions, PC-regular functions and Cauchy-subregular functions. Conditions under which PC-regular functions, strongly uniformly continuous functions, CBC-regular functions are stable under reciprocation, are also studied. Investigation on some properties of CBC-regular functions gives us characterizations of cofinally Bourbaki-complete metric spaces and finitely chainable spaces.

Unless mentioned otherwise, $\mathbb{R}$ and its non-empty subsets carry the usual distance metric and all metric spaces are considered to be infinite. In the monograph, we take one numbering for the Definitions, one for the Examples, one for the Remarks and another one for the Propositions, Lemmas, Theorems and Corollaries, each numbering being restricted to its own chapter.

Contents

Foreword vii

Introduction ix

1. Preliminaries 1

1.1 Variants of Complete Metric Spaces 1
1.2 Functions between Metric Spaces that Preserve Certain Sequences 7
1.3 Stronger Forms of Continuity on Metric Spaces 11
1.3.1 Lipschitz-type Functions 11
1.3.2 Strong Uniform Continuity 14

2. Cofinally Complete Metric Spaces 19

2.1 Cofinal Completeness vis-*à*-vis Uniform Paracompactness: Earlier Results 19
2.2 Cofinal Completeness vis-*à*-vis Local Compactness Functional 23
2.3 Cofinal Completeness vis-*à*-vis Functions between Metric Spaces 28

3. Cofinal Completions 49

3.1 Local Total Boundedness Functional 49
3.2 Cofinal Completion vis-*à*-vis Cauchy-regular Functions 52
3.3 Cauchy-subregular Functions 61
3.4 Some More Characterizations 67

4. Cofinal Completeness vis-*à*-vis Hyperspaces 75

4.1 Some Hyperspace Topologies 75
4.2 Cofinal Completeness vis-*à*-vis Hyperspaces 78
4.3 Cofinal Completeness of the Space $(AC(X), H_d)$ 88

5. Stronger Cofinal Completeness 97

5.1 UC Spaces 97
5.2 Cofinally Bourbaki-complete Metric Spaces 112

List of Symbols 125

Bibliography 127

Index 133

Chapter 1

Preliminaries

In this chapter, a brief description of some classes of metric spaces is given in which certain sequences, which are weaker than Cauchy sequences, have cluster points. In the first section, we compare different families of complete metric spaces, namely cofinally complete spaces, UC spaces, Bourbaki-complete spaces and cofinally Bourbaki-complete spaces. Then in the next section, we describe some functions between metric spaces that preserve the aforementioned weaker forms of Cauchy sequences. In the last section, some Lipschitz-type functions and strongly uniformly continuous functions are briefly discussed which are stronger than the continuous functions.

1.1 Variants of Complete Metric Spaces

In this section, we define certain classes of metric spaces that lie in between the compact spaces and the complete ones. To be precise, we briefly describe cofinally complete spaces, UC spaces, Bourbaki-complete spaces and cofinally Bourbaki-complete spaces.

Recall that in a metric space (X,d), a sequence is *Cauchy* if for each $\varepsilon > 0$, there exists a residual set of indices $\mathbb{N}_\varepsilon$ such that each pair of terms, the indices of which come from $\mathbb{N}_\varepsilon$ are within ε distance of each other and (X,d) is called complete provided every Cauchy sequence in the space clusters. In [Howes (1995)], Howes introduced a new class of sequences, which he called cofinally Cauchy, obtained by replacing *residual* by *cofinal* in the definition of Cauchy sequences. Thereby we get a generalization of Cauchy sequences, precisely defined as follows.

Definitions 1.1. A sequence (x_n) in a metric space (X,d) is called *cofinally Cauchy* if for each positive ε, there exists an infinite subset $\mathbb{N}_\varepsilon$ of $\mathbb{N}$ such that for each $n,\ j \in \mathbb{N}_\varepsilon$, we have $d(x_n,x_j) < \varepsilon$.

A metric space (X,d) is said to be *cofinally complete* if every cofinally Cauchy sequence in X clusters.

Since every Cauchy sequence is cofinally Cauchy, every cofinally complete metric space is complete. One can find various characterizations of cofinally complete metric spaces in [Beer (2008)]. In fact, the characterizations of these metric spaces are comprehensively studied in the next chapter. For that we would like to mention the following result, which says that every cofinally Cauchy sequence with no constant subsequence has a cofinally Cauchy subsequence of distinct terms.

Proposition 1.1. *([Beer (2008)]) Let (x_n) be a cofinally Cauchy sequence in a metric space (X,d) with no constant subsequence. Then there is a pairwise disjoint family $\{\mathbb{M}_j : j \in \mathbb{N}\}$ of infinite subsets of $\mathbb{N}$ such that*

(a) *if $\{i, l\} \subseteq \bigcup\{\mathbb{M}_j : j \in \mathbb{N}\}$ then $x_i \neq x_l$; and*
(b) *if $i \in \mathbb{M}_j$ and $l \in \mathbb{M}_j$ then $d(x_i,x_l) < \frac{1}{j}$.*

Proof. Suppose (x_n) has a Cauchy subsequence. Since (x_n) has no constant subsequence, we can assume that (x_n) has a Cauchy subsequence $(x_{n_k})_{k\in\mathbb{N}}$ of distinct terms. Thus, for each $j \in \mathbb{N}$, there exists $m_j \in \mathbb{N}$ such that $d(x_{n_k},x_{n_l}) < \frac{1}{j}$ for all $k,\ l \geq m_j$. Let $\mathbb{N}_o = \{n_k : k \in \mathbb{N}\}$. Then partition $\mathbb{N}_o$ into countably many infinite subsets $\{\mathbb{K}_j : j \in \mathbb{N}\}$. Now for each $j \in \mathbb{N}$, choose $\mathbb{M}_j = \{n_k \in \mathbb{K}_j : k \geq m_j\}$.

If (x_n) has no Cauchy subsequence, choose an infinite $\mathbb{M}_1 \subseteq \mathbb{N}$ such that $0 < d(x_i,x_l) < 1$ for all $i,\ l \in \mathbb{M}_1$. Since (x_n) has no Cauchy subsequence, the set $\{x_i : i \in \mathbb{M}_1\}$ cannot be totally bounded. By passing to an infinite subset of $\mathbb{M}_1$, we can find $\varepsilon_1 < \frac{1}{2}$ such that $\varepsilon_1 < d(x_i,x_l) < 1$ for all $i,\ l \in \mathbb{M}_1$. Now choose an infinite $\mathbb{M}_2 \subseteq \mathbb{N}$ such that $0 < d(x_i,x_l) < \varepsilon_1$ for all $i,\ l \in \mathbb{M}_2$. By construction $\{x_i : i \in \mathbb{M}_1\} \cap \{x_i : i \in \mathbb{M}_2\}$ consists of at most one point. Also, $\{x_i : i \in \mathbb{M}_2\}$ is not totally bounded, so by passing to an infinite subset of $\{x_i : i \in \mathbb{M}_2\}$ we can assume the two sets are disjoint and further that there exists $\varepsilon_2 < \frac{1}{3}$ such that $\varepsilon_2 < d(x_i,x_l) < \frac{1}{2}$ for all $i,\ l \in \mathbb{M}_2$. Choosing an infinite $\mathbb{M}_3 \subseteq \mathbb{N}$ such that $0 < d(x_i,x_l) < \varepsilon_2$ for all $i,\ l \in \mathbb{M}_3$, by deleting at most two indices from $\mathbb{M}_3$ we can assume $\{\{x_i : i \in \mathbb{M}_j\} : j = 1,2,3\}$ is a pairwise disjoint family. Continuing in this way inductively we produce $\{\mathbb{M}_j : j \in \mathbb{N}\}$ with the required properties. □

As a consequence of the previous result, it is enough to show the clustering of cofinally Cauchy sequences of *distinct* terms in order to prove a space to be cofinally complete.

Another natural generalization of a Cauchy sequence is pseudo-Cauchy sequence. These sequences play a significant role in characterizing UC spaces, which is another significant and better-known class of metric spaces that lies between the class of complete metric spaces and that of compact metric spaces. These spaces are also widely known as *Atsuji* spaces [Nagata (1950); Rainwater (1959); Waterhouse (1965); Wong (1972); Aggarwal and Kundu (2016); Beer (1985, 1986); Kundu and Jain (2006)]. Occasionally, they have been called *normal metric spaces* [Mrówka (1965)] and *Lebesgue metric spaces* [Nadler and West (1981); Romaguera and Antonino (1993)] in the literature.

Definition 1.2. A metric space (X,d) is called a *UC space* if every real-valued continuous function on (X,d) is uniformly continuous.

Definition 1.3. A sequence (x_n) in (X,d) is said to be *pseudo-Cauchy* if $\forall\ \varepsilon > 0$ and $\forall\ n \in \mathbb{N}$, there exist $k,\ j \in \mathbb{N}$, $k \neq j$ such that $k,\ j > n$ and $d(x_k, x_j) < \varepsilon$.

In [Toader (1978)], Toader has proved the following result.

Proposition 1.2. *[Toader (1978)] A metric space is UC if and only if every pseudo-Cauchy sequence of distinct points in the space clusters.*

In the sequential characterization of UC spaces given by Toader in terms of pseudo-Cauchy sequences, it is required that the sequences have distinct terms. For example, in the UC space of positive integers equipped with the Euclidean metric, the sequence $(1,1,2,2,3,3,\ldots)$ is a pseudo-Cauchy sequence without a cluster point. Unlike cofinally Cauchy sequences, pseudo-Cauchy sequences need not have a pseudo-Cauchy subsequence with distinct terms. Since every cofinally Cauchy sequence is pseudo-Cauchy, *every UC space is cofinally complete*. On the other hand, $\mathbb{R}$ with the usual metric is a cofinally complete space which is not UC. Besides the sequential characterization of UC spaces, there is another well-known characterization of UC spaces [Hueber (1981)] in terms of the *isolation functional* $I : X \to [0,\infty)$, defined by $I(x) = d(x, X \setminus \{x\})$, as follows.

Proposition 1.3. *[Hueber (1981)] A metric space (X,d) is UC if and only if every sequence (x_n) in X with $\lim_{n\to\infty} I(x_n) = 0$ clusters.*

It is well-known that a metric space (X,d) is totally bounded if and only if every sequence in X has a Cauchy subsequence. The following useful result characterizes totally bounded metric spaces in terms of pseudo-Cauchy and cofinally Cauchy sequences. The routine proof is omitted.

Proposition 1.4. *([Beer (2008)]) Let (X,d) be a metric space. Then the following assertions are equivalent.*
(a) The metric space (X,d) is totally bounded.

(b) Each sequence in X is cofinally Cauchy.
(c) Each sequence in X is pseudo-Cauchy.

In [Atsuji (1958)], finitely chainable metric spaces are introduced which are weaker than totally bounded metric spaces but stronger than bounded metric spaces. Furthermore, in the same paper, it has been shown that the finitely chainable metric spaces are precisely those metric spaces on which every real-valued uniformly continuous function is bounded. Also see [Njåstad (1965); O'Farrell (2004)]. The precise definition of a finitely chainable metric space follows.

Definitions 1.4. Let (X,d) be a metric space and ε be a positive number, then an ordered set of points $\{x_0,x_1,\ldots,x_m\}$ in X satisfying $d(x_{i-1},x_i) < \varepsilon$ for $i = 1,2,\ldots,m$, is said to be an *ε-chain of length m* from x_o to x_m. We call X *ε-chainable* if every two points in X can be joined by an ε-chain, and X is called *chainable* if X is ε-chainable for every $\varepsilon > 0$.

Let A be a subset of X. Then, A is said to be *finitely chainable* in X if for every $\varepsilon > 0$, there are finitely many points $p_1, p_2, \ldots, p_r$ in X and a positive integer m such that every point of A can be joined with some p_j, $1 \leq j \leq r$ by an ε-chain of length m.

Remark 1.1. These finitely chainable sets are also called *Bourbaki-bounded sets* in the literature, [Beer and Garrido (2014, 2015); Garrido and Meroño (2014)], because these sets were considered in the book [Bourbaki (1966)]. Note that these sets are called bounded sets in [Bourbaki (1966); Hejcman (1959)] where these sets are considered in the context of uniform spaces.

Clearly, $\mathbb{R}$ is chainable. In fact, every connected metric space is chainable. Observe that unlike total boundedness, finite chainability of a set A depends essentially on the underlying space (X,d), that is if Y is a subset of (X,d), then $A \subseteq Y$ is finitely chainable in X whenever it is finitely chainable in Y, but the converse need not hold. For example, any infinite bounded uniformly discrete set in a normed linear space is finitely chainable in the whole space but not in itself. A subset A of X is called *discrete* if $\forall\, x \in A$, $\exists\, \delta > 0$ such that $d(x,y) \geq \delta$ $\forall$ $y \in A \setminus \{x\}$, and it is called *uniformly discrete* if δ does not depend on x, that is, $\exists$ $\delta > 0$ such that $d(x,y) \geq \delta$ $\forall\, x,y \in A$, $x \neq y$.

Recently Garrido and Meroño have defined two new classes of sequences in [Garrido and Meroño (2014)], which they have called Bourbaki-Cauchy and cofinally Bourbaki-Cauchy. These sequences appeared when they were looking for sequential characterizations, of the Bourbaki-bounded sets, similar to that of total boundedness in terms of Cauchy sequences. The sequences are defined as follows:

Definitions 1.5. Let (X,d) be a metric space. A sequence (x_n) is said to be:

(a) *Bourbaki-Cauchy* in X if for every $\varepsilon > 0$, there exist $m \in \mathbb{N}$ and $n_o \in \mathbb{N}$ such that whenever $n > j \geq n_o$, the points x_j and x_n can be joined by an ε-chain of length m.

(b) *cofinally Bourbaki-Cauchy* in X if for every $\varepsilon > 0$, there exist $m \in \mathbb{N}$ and an infinite subset $\mathbb{N}_\varepsilon$ of $\mathbb{N}$ such that the points x_j and x_n can be joined by an ε-chain of length m for every $j,\ n \in \mathbb{N}_\varepsilon$.

Clearly every Cauchy sequence is Bourbaki-Cauchy and every Bourbaki-Cauchy sequence is cofinally Bourbaki-Cauchy. The aforesaid sequential characterizations of finitely chainable metric spaces, given by Garrido and Meroño, are as follows.

Proposition 1.5. *([Garrido and Meroño (2014)]) For a metric space (X,d) and $B \subseteq X$, the following statements are equivalent.*

(a) *The set B is finitely chainable in X.*

(b) *Every sequence in B has a Bourbaki-Cauchy subsequence in X.*

(c) *Every sequence in B is cofinally Bourbaki-Cauchy in X.*

Using the aforesaid two classes of sequences, Garrido and Meroño defined Bourbaki-complete and cofinally Bourbaki-complete metric spaces [Garrido and Meroño (2014)] naturally in the following manner.

Definitions 1.6. A metric space (X,d) is said to be *(cofinally) Bourbaki-complete* if every (cofinally) Bourbaki-Cauchy sequence in (X,d) clusters.

It is easy to see that a metric space (X,d) is complete if and only if every totally bounded set in it is relatively compact (that is, its closure in X is compact). Similarly, the Bourbaki complete spaces are precisely those in which each finitely chainable set is relatively compact [Garrido and Meroño (2014)]. Note that for proving a space to be (cofinally) Bourbaki-complete, it is enough to prove that every (cofinally) Bourbaki-Cauchy sequence of distinct terms in the space clusters. It is immediate that every (cofinally) Bourbaki-complete metric space is complete. Moreover, cofinal Bourbaki-completeness implies both Bourbaki-completeness and cofinal completeness. Let us note the following interesting result from [Garrido and Meroño (2014)] which says that every UC space is cofinally Bourbaki-complete.

Proposition 1.6. *([Garrido and Meroño (2014)]) Every UC space (X,d) is cofinally Bourbaki-complete.*

Proof. Let (x_n) be a cofinally Bourbaki-Cauchy sequence with no constant subsequence. Then for each $j \in \mathbb{N}$, there exist $m_j \in \mathbb{N}$ and an infinite subset $\mathbb{N}_j$ of $\mathbb{N}$ such that the points x_k and x_n can be joined by $\frac{1}{j}$-chain of length m_j for every k, $n \in \mathbb{N}_j$. Choose a subsequence $(x_{n_k})_{k\in\mathbb{N}}$ of (x_n) such that $n_j \in \mathbb{N}_j$ for every $j \in \mathbb{N}$. Then $I(x_{n_j}) < \frac{1}{j}$ and hence $\lim_{j\to\infty} I(x_{n_j}) = 0$. Since (X,d) is a UC space, the sequence $(x_{n_k})_{k\in\mathbb{N}}$ has a cluster point. Thus, (X,d) is cofinally Bourbaki-complete. □

The converse of the previous result need not hold, for example, consider $\mathbb{R}$ with the usual distance metric. In fact, many such examples can be constructed by noticing that every boundedly compact metric space (in which every closed and bounded subset is compact) is cofinally Bourbaki-complete but it need not be a UC space, for example consider the space $\{n, n+\frac{1}{n} : n \in \mathbb{N}\}$ with the usual distance metric. The following two examples show that there is no direct relation between cofinal completeness and Bourbaki-completeness.

Example 1.1. Let l_2 be the Hilbert space, and let $X = \bigcup_{n\in\mathbb{N}} A_n$, where

$$A_n = \{e_n\} \cup \left\{ e_n + \frac{1}{n} e_k : k \in \mathbb{N} \right\},$$

$\{e_n : n \in \mathbb{N}\}$ is the standard orthonormal basis of l_2. Then (X,d) is discrete and Bourbaki-complete, where 'd' is the metric induced by the l_2-norm, because every Bourbaki-Cauchy sequence in (X,d) is eventually constant. On the other hand, by enumerating X, we will get a cofinally Cauchy sequence which certainly does not cluster and hence (X,d) is not cofinally complete. Moreover, X is locally compact but it is not uniformly locally compact as for every $n \in \mathbb{N}$, the closed ball of radius $\frac{1}{n}$ and center e_n is the infinite discrete space A_n which is not compact. Note that a Hausdorff topological space (X,τ) is called *locally compact* if every point of X has a neighborhood whose closure is a compact subset of X, while a metric space (X,d) is called *uniformly locally compact* if there exists a $\delta > 0$ such that the closed ball $C(x,\delta)$ is compact for every $x \in X$.

Example 1.2. For an infinite cardinal number m, let $X = J(\mathrm{m})$ the hedgehog space of spininess m. That is, for a set S with cardinal m, the space X is the disjoint union $\bigcup_{s\in S}([0,1]\times\{s\})$ (after identifying all the points of the form $(0,s)$ to one only point $\mathbf{0} \equiv [(0,s)]$), endowed with the metric:

$$d([(x,s_1)],[(y,s_2)]) = \begin{cases} |x-y| & : s_1 = s_2 \\ x+y & : s_1 \neq s_2 \end{cases}$$

Then (X,d) is finitely chainable as for every $\varepsilon > 0$, there exists $m > \frac{1}{\varepsilon}$ such that every element of X can be joined with $\mathbf{0}$ by an ε-chain of length m. Moreover, (X,d) is not Bourbaki-complete, since the sequence $([(1,s_n)])_{n\in\mathbb{N}}$, where $s_n \neq s_m$ for all $n \neq m$, is Bourbaki-Cauchy, but it does not cluster. Now we claim that (X,d) is cofinally complete. Let $([(x_n,s_n)])_{n\in\mathbb{N}}$ be a cofinally Cauchy sequence in (X,d). Then for every $\varepsilon > 0$, there exists $\mathbb{N}_\varepsilon \subseteq \mathbb{N}$ such that $d([(x_i,s_i)],[(x_j,s_j)]) < \varepsilon$, for all $i,\ j \in \mathbb{N}_\varepsilon$. If for some $\varepsilon > 0$, there exists an infinite subset C of $\mathbb{N}_\varepsilon$ such that for every $i,\ j \in C$, we have $s_i = s_j$, then $([(x_n,s_n)])_{n\in\mathbb{N}}$ clusters because $[0,1]$ is compact. Otherwise, we can choose a subsequence $([(x_{n_k},s_{n_k})])_{k\in\mathbb{N}}$ of $([(x_n,s_n)])_{n\in\mathbb{N}}$ such that $d([(x_{n_k},s_{n_k})],\mathbf{0}) < \frac{1}{k}$ and hence the subsequence converges to $\mathbf{0}$.

Note that Example 1.1 shows that every Bourbaki-complete metric space need not be cofinally Bourbaki-complete and Example 1.2 shows that every cofinally complete space need not be cofinally Bourbaki-complete. The following implication diagram illustrates the relations among the spaces we consider in the upcoming chapters.

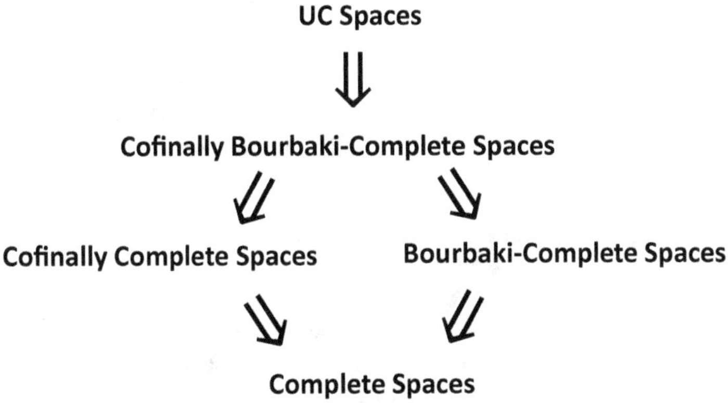

1.2 Functions between Metric Spaces that Preserve Certain Sequences

The significance of Cauchy sequences in the theory of metric spaces gave birth to a new class of functions that lies strictly in between the class of uniformly continuous functions and that of continuous functions, known as Cauchy-regular functions.

Definition 1.7. A function $f:(X,d)\to(Y,\rho)$ between two metric spaces is called *Cauchy-regular* if $(f(x_n))$ is Cauchy in (Y,ρ) for every Cauchy sequence (x_n) in (X,d).

Remark 1.2. Note that Cauchy-regular functions have also been widely known as *Cauchy-continuous* functions in the literature [Lowen-Colebunders (1989); Aggarwal and Kundu (2016); Beer and Garrido (2015, 2016)].

Many authors have investigated various properties and applications of Cauchy-regular functions [Aggarwal and Kundu (2016); Beer and Garrido (2015, 2016); Borsík (1988, 2000); Das *et al.* (2020); Jain and Kundu (2007); Snipes (1977, 1981)]. Cauchy-regular functions are important because: (1) some of the most useful theorems about uniformly continuous functions hold also for Cauchy-regular functions; and (2) many of the important functions which occur in analysis are Cauchy-regular but not uniformly continuous. Observe that every continuous function on a complete metric space is Cauchy-regular, whereas every Cauchy-regular function on a totally bounded metric space is uniformly continuous. In fact, every continuous function on a metric space (X,d) is Cauchy-regular if and only if (X,d) is complete; and every Cauchy-regular function on (X,d) is uniformly continuous if and only if the completion space $(\widehat{X},d)$ is UC [Beer (1986)].

After being motivated by the significant role that Cauchy-regular functions play in the theory of metric spaces, in 2017, Aggarwal and Kundu defined functions that preserve various weaker forms of Cauchy sequences [Aggarwal and Kundu (2017a)] that we have discussed in the previous section. Let us first record their definitions.

Definitions 1.8. A function $f:(X,d)\to(Y,\rho)$ between two metric spaces is said to be:

(*a*) *cofinally Cauchy regular* (or *CC-regular* for short) if $(f(x_n))$ is cofinally Cauchy in (Y,ρ) for every cofinally Cauchy sequence (x_n) in (X,d).
(*b*) *pseudo-Cauchy regular* (or *PC-regular* for short) if $(f(x_n))$ is pseudo-Cauchy in (Y,ρ) for every pseudo-Cauchy sequence (x_n) in (X,d).
(*c*) *Bourbaki-Cauchy regular* (or *BC-regular* for short) if $(f(x_n))$ is Bourbaki-Cauchy in (Y,ρ) for every Bourbaki-Cauchy sequence (x_n) in (X,d).
(*d*) *cofinally Bourbaki-Cauchy regular* (or *CBC-regular* for short) if $(f(x_n))$ is cofinally Bourbaki-Cauchy in (Y,ρ) for every cofinally Bourbaki-Cauchy sequencc (x_n) in (X,d).

Certainly, uniformly continuous functions preserve all the aforementioned sequences. Though these newly defined classes of functions are not contained in the class of continuous functions, they help in giving characterizations of various variants of complete metric spaces. We will see the details in the forthcoming chapters. In [Aggarwal and Kundu (2017a)], Aggarwal and Kundu studied the boundedness of these functions preserving certain sequences and proved that every real-valued CC-regular (PC-regular) function on a metric space (X,d) is bounded if and only if (X,d) is totally bounded, whereas the boundedness of BC-regular functions and CBC-regular functions characterize finitely chainable metric spaces. Further, in the same article, it was shown that there is no direct relation between any of these functions except for the following one.

Proposition 1.7. *[Aggarwal and Kundu (2016, 2017a)] Every PC-regular function between any two metric spaces is CC-regular.*

Proof. Let (x_n) be a cofinally Cauchy sequence in (X,d). Suppose that $(f(x_n))$ is not cofinally Cauchy. Then there exists $\varepsilon_o > 0$ such that for all $n \in \mathbb{N}$, $f(x_m) \in B(f(x_n), \varepsilon_o)$ for at most finitely many $m \in \mathbb{N}$. For $k = 1$, choose $m_1,\ n_1 \in \mathbb{N},\ m_1 < n_1$ such that $d(x_{m_1}, x_{n_1}) < 1$ and $\rho(f(x_{m_1}), f(x_{n_1})) \geq \varepsilon_o$. Similarly, for $k = 2$, choose $m_2,\ n_2 \in \mathbb{N},\ n_1 < m_2 < n_2$ such that $d(x_{m_2}, x_{n_2}) < \frac{1}{2}$ and $\rho(f(x_{v_p}), f(x_{w_q})) \geq \varepsilon_o$ for $v_p \neq w_q,\ \ p,\ q \in \{1,2\}$ and $v,\ w \in \{m,n\}$. By induction, we can choose $m_k,\ n_k \in \mathbb{N},\ \ n_{k-1} < m_k < n_k$ such that $d(x_{m_k}, x_{n_k}) < \frac{1}{k}$ and $\rho(f(x_{v_p}), f(x_{w_q})) \geq \varepsilon_o$ for $v_p \neq w_q,\ \ p,\ q \in \{1,\ldots,k\}$ and $v,\ w \in \{m,n\}$. Then the sequence $(x_{m_1}, x_{n_1}, x_{m_2}, x_{n_2}, \ldots)$ is pseudo-Cauchy in (X,d) but $(f(x_{m_1}), f(x_{n_1}), f(x_{m_2}), f(x_{n_2}), \ldots)$ is not pseudo-Cauchy in (Y,ρ), which is a contradiction. Thus, $(f(x_n))$ is cofinally Cauchy in (Y,ρ). □

Note that the converse of the previous proposition need not be true. For example, consider the function $f : \mathbb{R} \to \mathbb{R}$ defined as: $f(x) = x^2$. Then f is CC-regular (see Theorem 2.8), but f is not PC-regular as the image of the pseudo-Cauchy sequence $(1,\ 1+1,\ 2,\ 2+\frac{1}{2},\ 3,\ 3+\frac{1}{3},\ldots)$ under f is not pseudo-Cauchy. A CBC-regular function need not be BC-regular. For example, let $f : \mathbb{R} \to \{0,1\}$ be defined as:

$$f(x) = \begin{cases} 1: & x \in \mathbb{Q} \\ 0: & \text{otherwise} \end{cases}$$

Here note that if we would have taken the range of the function to be $\mathbb{R}$ instead of $\{0,1\}$, then this example would not have worked. Furthermore, a BC-regular function need not be CBC-regular: as a metric subspace of the Hilbert space ℓ_2, let $X = \{e_n + \frac{1}{n}e_k : n,\ k \in \mathbb{N}\}$. Let $\{x_n : n \in \mathbb{N}\}$ be an enumeration of the countable set X. Then the function defined as: $f : X \to \mathbb{N}$, $f(x_n) = n$, is BC-regular as every Bourbaki-Cauchy sequence in (X,d), where d is the metric induced by ℓ_2-norm, is eventually constant. But f is not CBC-regular as (x_n) is cofinally Cauchy in (X,d) and its image under f is not cofinally Bourbaki-Cauchy in $\mathbb{N}$. Also, a CBC-regular function need not be CC-regular. For example, let $f:\{\frac{1}{n}:n\in\mathbb{N}\}\to\ell_2$ be defined as: $f(\frac{1}{n})=e_n$. In fact, the converse also need not be true: let $f:\ell_2\to\mathbb{N}$ be defined as,

$$f(x)=\begin{cases} n: \ x=e_n \text{ for some } n\in\mathbb{N} \\ 1: \text{otherwise} \end{cases}$$

We will study more properties of these functions along with their relations with each other in the forthcoming chapters. Let us now discuss about some extension results based on continuous functions, Cauchy-regular functions, and uniformly continuous functions. We first recall the following version of the Tietze Extension Theorem.

Theorem 1.1. *[Dugundji (1966), p. 149] Let A be a closed subset of a metric space* (X,d). *If* $f:A\to(a,b)$ *is a continuous function, where* $\{a,b\}\subseteq\mathbb{R}$, *then there exists a continuous function* $g:X\to(a,b)$ *which extends* f, *that is,* $g|_A=f$.

The next result describes the significance of Cauchy-regular functions. Its routine proof is omitted.

Proposition 1.8. *([Snipes (1977)]) Let A be a subset of a metric space* (X,d) *and* (Y,ρ) *be a complete metric space. If* $f:(A,d)\to(Y,\rho)$ *is Cauchy-regular, then there exists a unique continuous function* $\tilde{f}:\overline{A}\to Y$ *defined on the closure* $\overline{A}$ *of A which extends f, that is,* $\tilde{f}(x)=f(x)$ *for all* $x\in A$. *Moreover, the function* $\tilde{f}$ *is Cauchy-regular.*

The next two results can be easily proved by using Proposition 1.8, Tietze's extension theorem and the completion of the metric space (X,d).

Corollary 1.1. *Let A be a subset of a metric space* (X,d) *and* $f:(A,d)\to\mathbb{R}$ *be a Cauchy-regular function. Then there exists a Cauchy-regular function* $g:(X,d)\to\mathbb{R}$ *such that* $g|_A=f$.

Corollary 1.2. *([Borsík (1988)]) Let A be a subset of a metric space* (X,d) *and* $f:(A,d)\to[0,1]$ *be a Cauchy-regular function. Then there exists a Cauchy-regular function* $g:(X,d)\to[0,1]$ *such that* $g|_A=f$.

Let us also recall the famous McShane's extension theorem for uniformly continuous functions.

Theorem 1.2. *[McShane (1934)] Let A be a non-empty subset of a metric space* (X,d). *If f is a bounded real-valued uniformly continuous function on A, then* *there exists a uniformly continuous function $g : X \to \mathbb{R}$ which extends f, that is, $g|_A = f$.*

1.3 Stronger Forms of Continuity on Metric Spaces

We all are familiar with some stronger versions of continuous functions, like uniformly continuous functions, Cauchy-regular functions. Now we discuss in brief a few more interesting classes of functions which lies in the bigger class of continuous functions. These functions will then be used in upcoming chapters for giving various characterizations.

1.3.1 *Lipschitz-type Functions*

In analysis, there is another well-known form of continuity which is even stronger than uniform continuity, namely Lipschitz continuity. In [Beer *et al.* (2020); Beer and Garrido (2014, 2015, 2016, 2020); Luukkainen (1979); Garrido and Jaramillo (2008)], various authors considered some Lipschitz-type functions, namely locally Lipschitz, Cauchy- Lipschitz, uniformly locally Lipschitz, and Lipschitz in the small. In [Beer and Garrido (2014)], boundedness of these Lipschitz-type functions was characterized; in [Beer and Garrido (2015, 2016)], one can find applications of these functions to characterize UC spaces and cofinally complete metric spaces and recently in [Beer *et al.* (2020)], the authors discussed the conditions under which such (non-zero) functions are stable under reciprocation. Subsequently, in the forthcoming chapters, these Lipschitz-type functions are used to characterize different families of cofinally complete metric spaces, totally bounded metric spaces and finitely chainable metric spaces. Let us here recall some basic and interesting properties of these functions.

Definitions 1.9. A function $f : (X,d) \to (Y,\rho)$ between two metric spaces is said to be:

- (a) *Lipschitz* if there exists $k > 0$ such that $\rho(f(x), f(x')) \leq kd(x,x')$ for all $x,x' \in X$.
- (b) *Lipschitz in the small* if there exist $\delta > 0$ and $k > 0$ such that $\rho(f(x), f(x')) \leq kd(x,x')$, whenever $d(x,x') < \delta$.
- (c) *uniformly locally Lipschitz* if there exists $\delta > 0$ such that for every $x \in X$, there exists $k_x > 0$ such that $\rho(f(u), f(w)) \leq k_x d(u,w)$, whenever $u,w \in B(x,\delta)$.
- (d) *Cauchy-Lipschitz* if f is Lipschitz when restricted to the range of each Cauchy sequence (x_n) in X.

(e) *locally Lipschitz* if for each $x \in X$, there exists $\delta_x > 0$ such that f restricted to $B(x, \delta_x)$ is Lipschitz.

It is immediate that every Lipschitz in the small function between two metric spaces is uniformly locally Lipschitz. Also see [Leung and Tang (2017)]. Moreover, it was shown in [Beer and Garrido (2016)] that the collection of Cauchy-Lipschitz functions is contained in the class of locally Lipschitz functions whereas the collection itself contains the uniformly locally Lipschitz functions. But the reverse implications are not in general true. For example, $f : (0,\infty) \to (0,\infty)$ defined by: $f(x) = \frac{1}{x}$ is locally Lipschitz as for any $x \in (0,\infty)$, choose $\delta_x = \frac{x}{4}$ and then $k_x = \frac{16}{9x^2}$ would do the job, but f is not Lipschitz when restricted to $\{\frac{1}{n} : n \in \mathbb{N}\}$. Clearly, $f : \mathbb{R} \to \mathbb{R}$ defined by $f(x) = x^2$ is uniformly locally Lipschitz, but not Lipschitz in the small. For a function which is Cauchy-Lipschitz but not uniformly locally Lipschitz, consider the space $X = \{e_n + \frac{1}{n}e_k : n,\ k \in \mathbb{N}\}$ as a metric subspace of the Hilbert space ℓ_2. Evidently, every Cauchy sequence in (X,d) is eventually constant and thus has finite range. As a consequence, each function on X is Lipschitz when restricted to Cauchy sequences. Now consider the function $f : X \to \mathbb{R}$ defined by: $f(e_n + \frac{1}{n}e_k) = k$ for $n,\ k \in \mathbb{N}$. Then each $A_n = \{e_n + \frac{1}{n}e_k : k \in \mathbb{N}\}$ is of diameter $\frac{\sqrt{2}}{n}$ on which f so restricted is unbounded. Hence f cannot be uniformly locally Lipschitz.

Recall that every continuous function on a complete space is Cauchy-regular. The next result says that this analogue is maintained when we pass to the locally Lipschitz and Cauchy-Lipschitz functions.

Theorem 1.3. *([Beer and Garrido (2016)]) Let (X,d) be a metric space. Then the following statements are equivalent:*

(a) *The metric space (X,d) is complete.*
(b) *Each locally Lipschitz function on (X,d) with values in a metric space (Y,ρ) is Cauchy-Lipschitz.*
(c) *Each real-valued locally Lipschitz function on (X,d) is Cauchy-Lipschitz.*

Proof. The implication $(a) \Rightarrow (b)$ easily follows from the fact that a locally Lipschitz function is Lipschitz when restricted to each convergent sequence, whereas the implication $(b) \Rightarrow (c)$ is immediate.

$(c) \Rightarrow (a)$: Let (x_n) be a non-convergent Cauchy sequence in (X,d). Let (x_n) converge to p in the completion $(\widehat{X},d)$ of (X,d). Consider the function, $f : X \to \mathbb{R}$ defined by: $f(x) = \frac{1}{d(x,p)}$. Then f fails to be bounded on $\{x_n : n \in \mathbb{N}\}$ and so f fails to be Cauchy-Lipschitz. However, f is locally Lipschitz being the composition of two locally Lipschitz functions, $x \mapsto d(x,p)$ and $\alpha \mapsto \frac{1}{\alpha}$ on $(0,\infty)$. □

It is easy to see that the collections of locally Lipschitz, Cauchy-Lipschitz and Lipschitz in the small functions are properly contained in the collections of continuous, Cauchy-regular and uniformly continuous functions respectively. Furthermore, it is noteworthy that these collections are not only contained but they are uniformly dense in the corresponding bigger collections with some restrictions on the codomain. The precise statements of these results are as follows.

Theorem 1.4. *([Miculescu (2000/01); Garrido and Jaramillo (2004); Beer and Garrido (2016)]) Let* (X,d) *be a metric space and* $(Y,\|\,.\,\|)$ *be a normed linear space. Then each continuous function* f *from* X *to* Y *can be uniformly approximated by locally Lipschitz functions, that is, given* $\varepsilon > 0$, *there exists a locally Lipschitz function* $g : (X,d) \to (Y,\|\,.\,\|)$ *with* $\sup_{x\in X} \|f(x) - g(x)\| \leq \varepsilon$.

Remark: The proof of Theorem 1.4 has been omitted as it is a bit complicated. But the interested reader may refer to the above mentioned references for the details.

Theorem 1.5. *([Beer and Garrido (2016)]) Let* (X,d) *be a metric space and* $(Y,\|\,.\,\|)$ *be a Banach space. Then each Cauchy-regular function* f *from* X *to* Y *can be uniformly approximated by Cauchy-Lipschitz functions.*

Proof. Let $f : X \to Y$ be a Cauchy-regular function and let $\varepsilon > 0$. Then by Proposition 1.8, f can be extended to a continuous function g from $\widehat{X}$ to Y. Subsequently, by Theorem 1.4, there exists a locally Lipschitz function $h : \widehat{X} \to Y$ such that $\sup_{x\in\widehat{X}} \|g(x) - h(x)\| \leq \varepsilon$. Since $(\widehat{X},d)$ is complete, Theorem 1.3 implies h is Cauchy-Lipschitz. Hence h restricted to X is the required function. □

Theorem 1.6. *([Beer and Garrido (2015); Garrido and Jaramillo (2008)]) Let* (X,d) *be a metric space and* $f : (X,d) \to \mathbb{R}$ *be a uniformly continuous function. Given* $\varepsilon > 0$, *there exists a Lipschitz in the small function* $g : (X,d) \to \mathbb{R}$ *with* $\sup_{x\in X} |f(x) - g(x)| \leq \varepsilon$.

Proof. Choose $\delta > 0$ such that $|f(x) - f(w)| < \varepsilon$ whenever $d(x,w) < \delta$. Choose $k \in \mathbb{N}$ such that $\frac{k\delta}{2} > 1 + \varepsilon$ ($\bigstar$).

Define $g : (X,d) \to \mathbb{R}$ by: $g(x) = \inf_{w\in B(x,\delta)}(f(w) + kd(x,w))$. Then for each $x \in X$, $f(x) \geq g(x) \geq \inf_{w\in B(x,\delta)} f(w) \geq f(x) - \varepsilon$.

Now we claim that g is Lipschitz in the small. We show that if $d(x_1,x_2) < \frac{\delta}{2}$, then $|g(x_2) - g(x_1)| \leq kd(x_2,x_1)$. Let $\lambda \in (0,1)$. By the definition of infimum, choose $w_1 \in B(x_1,\delta)$ such that $f(w_1) + kd(x_1,w_1) < g(x_1) + \lambda$. If $d(x_1,w_1) \geq \frac{\delta}{2}$, then by ($\bigstar$),

$$f(w_1) + kd(x_1,w_1) > f(x_1) - \varepsilon + 1 + \varepsilon > g(x_1) + \lambda,$$

which is impossible. So $d(x_1,w_1) < \frac{\delta}{2}$, which implies $d(x_2,w_1) < \delta$. Consequently,

$$g(x_2) \leq f(w_1)+kd(x_2,w_1) = f(w_1)+kd(x_1,w_1)+kd(x_2,w_1)-kd(x_1,w_1)$$
$$< g(x_1)+\lambda+kd(x_2,x_1)$$

Since $\lambda \in (0,1)$ was arbitrary, the proof is complete. □

1.3.2 *Strong Uniform Continuity*

We know that if $f:(X,d) \to (Y,\rho)$ is a continuous function and K is a compact subset of X, then f is uniformly continuous on K. In [Beer and Levi (2009a)], it was observed that something more than the uniform continuity of f on K would be true, which gave rise to an interesting concept called strong uniform continuity. More precisely, in such a case, f will be strongly uniformly continuous on K.

Definition 1.10. Let (X,d) and (Y,ρ) be metric spaces and B be a subset of X. A function $f:X \to Y$ is said to be *strongly uniformly continuous* on B if for all $\varepsilon > 0$, there exists $\delta > 0$ such that if $d(x,y) < \delta$ and $\{x,y\} \cap B \neq \emptyset$, then $\rho(f(x),f(y)) < \varepsilon$.

Note that a function f on a metric space (X,d) is continuous if and only if it is strongly uniformly continuous on $\{x\}$ for each $x \in X$. In this monograph, we also study the families of subsets of cofinally complete metric spaces on which every continuous function must be strongly uniformly continuous. We also investigate the condition under which every non-vanishing strongly uniformly continuous function is stable under reciprocation. One of the key tools which played an important role in the analysis of strong uniform continuity in [Beer and Levi (2009a)] was those functions between metric spaces which preserved totally bounded sets. For example, the authors in [Beer and Levi (2009a)] proved that if f is a function between two metric spaces which preserves totally bounded sets, then f is uniformly continuous if and only if the set of subsets on which f is strongly uniformly continuous is same as the set of subsets on which f is uniformly continuous. Actually, these functions preserving totally bounded sets, which were earlier mentioned in [Beer and Levi (2009b)], can be characterized in another familiar way through Cauchy-subregular functions. So let us first define Cauchy-subregular functions, which were first explicitly considered in [Gupta and Kundu (2022)].

Definition 1.11. A function $f:(X,d) \to (Y,\rho)$ between two metric spaces is said to be *Cauchy-subregular* if it takes Cauchy sequences to the sequences having a Cauchy subsequence.

Proposition 1.9. *[Beer and Levi (2009b)] Let $f : (X,d) \to (Y,\rho)$ be a function between two metric spaces. Then f preserves totally bounded sets if and only if f is Cauchy-subregular.*

It is clear that Cauchy-subregularity is a weaker form of Cauchy-regularity. We have seen that Cauchy-regular functions play an important role in characterizing complete metric spaces and their variants. In this monograph, we see that this weaker notion of Cauchy-subregularity can be profitably used to characterize such spaces. We investigate the relation of Cauchy-subregular functions with CC-regular and PC-regular functions and characterize those spaces on which every function that is both continuous and Cauchy-subregular is uniformly continuous. We also study answers to some questions like under what conditions Cauchy-subregularity implies Cauchy-regularity or uniform continuity?

Exercises

Exercise 1.1

Give examples of:

(a) cofinally Cauchy sequence which is not Cauchy.
(b) pseudo-Cauchy sequence which is not cofinally Cauchy.
(c) Bourbaki-Cauchy sequence which is not Cauchy.
(d) cofinally Bourbaki-Cauchy sequence which is not Bourbaki-Cauchy.

Exercise 1.2

Prove Proposition 1.2 and Proposition 1.3.

Exercise 1.3

Give examples of:

(a) complete space which is not cofinally complete.
(b) UC space which is not compact.

Exercise 1.4

[Kundu and Jain (2006)] Show that if (X,d) is a UC space then the set X' of all accumulation points of X is compact in (X,d). Also, think of an example of a complete metric space (X,d) such that X' is compact in X but (X,d) is not UC.

Exercise 1.5

Show that every connected metric space is chainable.

Exercise 1.6

Prove Proposition 1.5, Proposition 1.8 and Proposition 1.9.

Exercise 1.7

[Garrido and Meroño (2014)] Show that a metric space (X,d) is Bourbaki-complete if and only if every finitely chainable set in X is relatively compact.

Exercise 1.8

[Garrido and Meroño (2014)] Show that the following statements are equivalent.

(a) (X,d) is compact.
(b) (X,d) is totally bounded and cofinally complete.
(c) (X,d) is finitely chainable and cofinally Bourbaki-complete.

Exercise 1.9

[Beer and Garrido (2016)] Show that every uniformly locally Lipschitz function is Cauchy Lipschitz and every Cauchy Lipschitz function is locally Lipschitz.

Exercise 1.10

[Beer and Garrido (2015)] Let (X,d) be a metric space and let A be a non-empty subset of X. Prove that the following statements are equivalent.

(a) A is finite if and only if whenever (Y,ρ) is a metric space and $f:(X,d)\to(Y,\rho)$ is continuous, then $f|_A$ is Lipschitz.
(b) A is relatively compact if and only if whenever (Y,ρ) is a metric space and $f:(X,d)\to(Y,\rho)$ is locally Lipschitz, then $f|_A$ is Lipschitz.
(c) A is totally bounded if and only if whenever (Y,ρ) is a metric space and $f:(X,d)\to(Y,\rho)$ is uniformly locally Lipschitz, then $f|_A$ is Lipschitz.

Exercise 1.11

[Gupta and Kundu (2022)] Let $f:(X,d)\to(Y,\rho)$ be a function between two metric spaces. Prove that the following are equivalent.

(a) f is Cauchy-subregular.
(b) Whenever (Z,μ) is a metric space and $g:(Y,\rho)\to(Z,\mu)$ is a Cauchy-subregular function, then $g\circ f$ is Cauchy-subregular.

(c) Whenever (Z,μ) is a metric space and $g:(Y,\rho)\to(Z,\mu)$ is a Cauchy-regular function, then $g\circ f$ is Cauchy-subregular.

(d) Whenever (Z,μ) is a metric space and $g:(Y,\rho)\to(Z,\mu)$ is a Cauchy-Lipschitz function, then $g\circ f$ is Cauchy-subregular.

(e) If $g:(Y,\rho)\to\mathbb{R}$ is a uniformly locally Lipschitz function, then $g\circ f$ is Cauchy-subregular.

Chapter 2

Cofinally Complete Metric Spaces

In this chapter, we study cofinally complete metric spaces in detail. In order to make this monograph self-contained, the first section is dedicated to the study of various characterizations of such spaces in the literature. In the second section, the main focus is on equivalent conditions based on a geometric functional known as local compactness functional. In the last section, we emphasize on CC-regular functions and their relations with other functions, which leads us to explore the cofinally complete metric spaces to further extent.

2.1 Cofinal Completeness vis-*à*-vis Uniform Paracompactness: Earlier Results

The concept of completeness occupies a central role in the theory of metric spaces. Cofinal completeness constitutes an interesting stronger form of completeness that arose in the study of paracompactness in uniform spaces. In fact, in [Corson (1958)] cofinal completeness was considered first implicitly by Corson who was actually interested in some convenient formulation of paracompactness in terms of uniform structures. Recall that a Hausdorff space (X,τ) is *paracompact* if each open cover of X has an open locally finite refinement.

Let $\mathscr{D}$ be a diagonal uniformity on a set X. Then recall that a filter $\mathscr{F}$ in the uniform space $(X,\mathscr{D})$ is said to be *Cauchy* if for every entourage D, there is an $A \in \mathscr{F}$ such that $A \times A \subseteq D$. During the study, Corson defined another type of filters which he called weakly Cauchy as they were weaker than Cauchy filters. The weakly Cauchy filters are defined as follows.

Definition 2.1. A filter $\mathscr{F}$ in a uniform space $(X,\mathscr{D})$ is *weakly Cauchy* if for every $U \in \mathscr{D}$, there exist a filter $\mathscr{H}_U$ and $H \in \mathscr{H}_U$ such that $\mathscr{F} \subseteq \mathscr{H}_U$ and $H \times H \subseteq U$.

Here is the result of Corson.

Theorem 2.1. *([Corson (1958)]) For a Hausdorff topological space X, the following statements are equivalent:*

(*a*) *X is paracompact.*
(*b*) *The set of all neighborhoods of the diagonal is a uniformity for X, and the product of X with every compact space is normal.*
(*c*) *There is a uniformity for X under which every weakly Cauchy filter has a cluster point.*
(*d*) *If $\mathscr{F}$ is a filter in X such that the image of $\mathscr{F}$ has a cluster point in any metric space into which X is continuously mapped, then $\mathscr{F}$ has a cluster point in X.*

Note that a filter $\mathscr{F}$ in a uniform space (X,μ), where μ is a covering uniformity on X, is said to be *weakly Cauchy* if for each $\mathfrak{U} \in \mu$ there is a filter $\mathscr{G}$ containing $\mathscr{F}$ and a $G \in \mathscr{G}$ such that $G \subseteq U$ for some $U \in \mathfrak{U}$ (or equivalently, for each $\mathfrak{U} \in \mu$, there exists a $U \in \mathfrak{U}$ such that $U \cap F \neq \emptyset$ for each $F \in \mathscr{F}$). In [Howes (1971)], such equivalent formulation of weakly Cauchy filters was utilized by Howes for giving a necessary and sufficient condition for a uniform space (X,μ) to be cofinally complete. Howes considered cofinal completeness in terms of nets and covering uniformities. Now, we give the precise definitions of cofinally Cauchy nets and cofinally complete uniform spaces.

Definitions 2.2. A net $(x_\lambda)_{\lambda \in \Lambda}$ in a uniform space (X,μ) is said to be *cofinally Cauchy* if for each $\mathfrak{U} \in \mu$ there exists $U \in \mathfrak{U}$ such that the net is cofinally contained in U, that is, for every $\lambda \in \Lambda$ there exists $\lambda_o \geq \lambda$ such that $x_{\lambda_o} \in U$.

A uniform space (X,μ) is called *cofinally complete* if every cofinally Cauchy net has a cluster point.

The next result was given by Howes which stated the relation between cofinal completeness and weakly Cauchy filters. As far as the proof is concerned, notice that the net based on a weakly Cauchy filter is cofinally Cauchy, while the filter generated by a cofinally Cauchy net is weakly Cauchy.

Theorem 2.2. *([Howes (1971)]) A uniform space (X,μ) is cofinally complete if and only if each weakly Cauchy filter in (X,μ) clusters.*

Remark 2.1. By Theorem 2.1 and Theorem 2.2, it can be concluded that a Hausdorff space X is paracompact if and only if it admits a compatible uniformity μ such that (X,μ) is cofinally complete.

Furthermore, in [Howes (1971)] it was shown that the completion of a preparacompact uniform space is cofinally complete, where a uniform space (X,μ) is said to be *preparacompact* if each cofinally Cauchy net has a Cauchy subnet. In fact, in [Howes (1995)] (page 107, Theorem 4.12) Howes proved that a uniform space is preparacompact if and only if its completion is cofinally complete (we prove this in the setting of metric spaces in the next chapter in Theorem 3.3). Surprisingly, the investigation of cofinal completeness in metric spaces started from [Howes (1995)]. Besides this, Howes introduced a new type of spaces in [Howes (1995)], which he called regularly bounded, in order to consider the conditions on a metric space under which cofinal completeness and completeness become identical. A metric space is said to be *regularly bounded* if every closed and bounded subset is totally bounded. Thus, a metric space is regularly bounded if and only if each bounded sequence has a Cauchy subsequence. As a consequence, now it is easy to see that in a regularly bounded metric space, completeness and cofinal completeness are equivalent.

It is well-known that every metric space is paracompact. Interestingly, in [Rice (1977)] for a Hausdorff uniform space (X,μ), Rice introduced the concept of *uniform paracompactness*: every open cover has a uniformly locally finite open refinement. Note that a refinement $\mathscr{V}$ of a cover $\mathscr{U}$ is said to be *uniformly locally finite* if there exists a uniform cover $\mathfrak{W}$ such that each member of $\mathfrak{W}$ meets only finitely many members of $\mathscr{V}$. Rice gave some characterizations of uniformly paracompact metric spaces which we mention in our next result without proof.

Theorem 2.3. *([Rice (1977)]) For a uniform space (X,μ), the following statements are equivalent:*

(a) *The uniform space (X,μ) is uniformly paracompact.*
(b) *If $\mathscr{U}$ is an open cover, then $\mathscr{U}_f$ (the cover consisting of finite unions of the members of $\mathscr{U}$) is a uniform cover.*
(c) *If $\mathscr{U}$ is an open cover, then there exists a uniform cover $\mathfrak{V}$ such that $\mathscr{U}/V$ has a finite subcover for each $V \in \mathfrak{V}$, where $\mathscr{U}/V = \{V \cap U : U \in \mathscr{U}\}$.*

The following corollary can be viewed as an equivalent formulation of the previous theorem in terms of metric spaces and hence we get the corresponding definition of uniform paracompactness in metric spaces.

Corollary 2.1. *The following statements are equivalent for a metric space (X,d):*

(a) *For every open cover $\mathscr{U}$ of X, there is an open refinement $\mathscr{V}$ and $\delta > 0$ such that for each $x \in X$, $B(x,\delta)$ intersects at most finitely many members of $\mathscr{V}$.*
(b) *For every open cover $\mathscr{U}$ of X, there exists $\delta > 0$ such that each subset of X with diameter less than δ admits a finite subcover from $\mathscr{U}$.*

In the same paper, Rice proved that a locally compact uniform space is uniformly paracompact if and only if it is uniformly locally compact, where a uniform space (X,μ) is said to be *uniformly locally compact* if there exists a $\mathfrak{U} \in \mu$ consisting of compact sets (the result is proved in the context of metric spaces in Theorem 2.8). Note that the hedgehog space based on infinitely many copies of the unit interval (see Example 1.2) is uniformly paracompact, but it is not locally compact.

Subsequently, the reviewer of Rice's paper for Mathematical Reviews, observed in [Smith (1978)] that uniform paracompactness is equivalent to net cofinal completeness for a Hausdorff uniform space.

Theorem 2.4. *([Howes (1995)], page 95, Theorem 4.6) A uniform space is cofinally complete if and only if it is uniformly paracompact.*

A few years later in [Hohti (1981)], with an aim to give a "sufficiently clear characterization" of uniformly paracompact metric spaces, Hohti gave the following result.

Theorem 2.5. *([Hohti (1981)]) A necessary and sufficient condition for a metric space (X,d) to be uniformly paracompact is that there exists a compact set $K \subseteq X$ such that $X \setminus B(K,\varepsilon)$ is uniformly locally compact for all $\varepsilon > 0$.*

Remark 2.2. A similar characterization of uniformly paracompact spaces (and equivalently, of cofinally complete metric spaces) is proved in the next section in Theorem 2.7.

In order to characterize metrizable spaces which admit a cofinally complete metric, Romaguera introduced the notion of a cofinally Čech complete space [Romaguera (1998)], where a Tychonoff space X is called *cofinally Čech complete* if there is a countable collection $\{\mathscr{G}_n : n \in \mathbb{N}\}$ of open covers of X satisfying the property that whenever $\mathscr{F}$ is a filter on X such that for each $n \in \mathbb{N}$ there is some $G_n \in \mathscr{G}_n$ which meets all the members of $\mathscr{F}$, then $\mathscr{F}$ has a cluster point. He showed that *a metrizable space admits a cofinally complete metric if and only if it is cofinally Čech complete*, and then obtained as an application that the space has a compatible cofinally complete metric if and only if the set of points that admit no compact neighborhood is compact (necessity was first observed in [Rice (1977)]). After ten years, Beer gave a short and direct proof of the Romaguera's theorem on compactness of the set of points that admit no compact neighborhood, which he denoted by $nlc(X)$.

Theorem 2.6. *([Beer (2008)]) Let X be a metrizable topological space. Then the following conditions are equivalent:*

(a) *X has a compatible cofinally complete metric.*
(b) *The set $nlc(X)$ is compact.*
(c) *Whenever F is a closed subset of $nlc(X)$, F has a countable base for its neighborhood system.*

Remark 2.3. As a consequence of the previous result, it can be concluded that every locally compact metrizable space admits a compatible cofinally complete metric. Moreover, the Hilbert space l_2 does not admit any compatible cofinally complete metric.

It is noteworthy that Beer gave a new direction to the study of cofinal completeness in metric spaces in [Beer (2008)] by adding various nice and useful characterizations of cofinally complete metric spaces which we discuss in our next section.

2.2 Cofinal Completeness vis-*à*-vis Local Compactness Functional

Beer cast a new light on those metric spaces in which every cofinally Cauchy sequence has a cluster point, widely known as cofinally complete metric spaces [Beer (2008)], where cofinally Cauchy sequence is a natural generalization of the well-known Cauchy sequence. More precisely, a sequence in a metric space is called cofinally Cauchy if for each positive ε, there exists a cofinal (rather than residual) set of indices whose corresponding terms are ε-close. In this section, we discuss about the various characterizations of cofinally complete metric spaces given by Beer along with many other useful characterizations.

First let us mention some notational conventions and required definitions. If x is a point in a metric space (X,d) and A is a nonempty subset of X, then $d(x,A) = \inf\{d(x,a) : a \in A\}$, whereas if $A = \emptyset$, then $d(x,A) = \infty$. The open ε-enlargement of A is denoted by A^{ε}, that is, $A^{\varepsilon} = \bigcup_{x\in A} B(x,\varepsilon)$. If A and B are non-empty subsets of X, then the set $D(A,B)$(or $D_d(A,B)) = \inf\{d(a,b) : a \in A, b \in B\}$ is called the *gap* between A and B. The Hausdorff distance between A and B is defined as:

$$\begin{aligned} H_d(A,B) &= \max\{\sup\{d(a,B) : a \in A\},\ \sup\{d(b,A) : b \in B\}\} \\ &= \inf\{\varepsilon > 0 : B \subseteq A^{\varepsilon} \text{ and } A \subseteq B^{\varepsilon}\} \end{aligned}$$

It is well-known that complete metric spaces are characterized by the non-empty intersection of decreasing sequences of non-empty closed sets $\langle F_n \rangle$ having either of the following properties: (1) $\lim_{n\to\infty} diam(F_n) = 0$, or (2) $\lim_{n\to\infty} \alpha(F_n) = 0$, where for each $A \subseteq X$

$$\alpha(A) = \inf\{\varepsilon > 0 : A \text{ is contained in a finite union of sets of diameter} < \varepsilon\}.$$

In the literature, the functional α is called the *Kuratowski measure of non-compactness*. Similar characterizations are given in [Beer (2008)] for cofinally complete metric spaces in terms of the following functionals:

$$\overline{\nu}(A) = \sup\{\nu(x) : x \in A\} \text{ and}$$
$$\underline{\nu}(A) = \inf\{\nu(x) : x \in A\},$$

where the geometric functional $\nu(.)$ is defined as follows: let $x \in X$. If x has a compact neighbourhood, set $\nu(x) = \sup\{\varepsilon > 0 : C(x,\varepsilon) \text{ is compact}\}$, otherwise, put $\nu(x) = 0$. Alternatively,

$$\nu(x) = \inf\{\varepsilon > 0 : C(x,\varepsilon) \text{ is non-compact}\}$$

This geometric functional is called the *local compactness functional on* X. The set $\{x \in X : \nu(x) = 0\}$ is the set of points of non-local compactness of X, which is denoted by $nlc(X)$. Thus a metric space (X,d) is said to be locally compact if $\nu(x) > 0 \; \forall \, x \in X$, whereas it is called uniformly locally compact if $\inf\{\nu(x) : x \in X\} > 0$.

The next lemma is an interesting observation regarding the functional.

Lemma 2.1. *([Beer (2008)]) Let (X,d) be a metric space. If X is not boundedly compact, then $\nu : X \to [0,\infty)$ is uniformly continuous.*

Proof. Let $\varepsilon > 0$ and $x, y \in X$ such that $d(x,y) \leq \frac{\varepsilon}{2}$. We claim that $\nu(x) \leq \nu(y) + \varepsilon$. Suppose $\nu(x) > \nu(y) + \varepsilon$. Then there exists $r > \nu(y) + \varepsilon$ such that $C(x,r)$ is compact. Since $C(y, r - \frac{\varepsilon}{2}) \subseteq C(x,r)$, $\nu(y) \geq r - \frac{\varepsilon}{2}$. This implies $r > r + \frac{\varepsilon}{2}$, which is a contradiction. Thus, $\nu(x) \leq \nu(y) + \varepsilon$. By interchanging the roles of x and y, we get $|\nu(x) - \nu(y)| \leq \varepsilon$. □

Let us now discuss about almost nowhere locally compact subsets. These sets were first defined in [Beer and Di Maio (2012)] in order to give the following characterization of cofinally complete metric spaces: a metric space (X,d) is cofinally complete if and only if whenever A is a closed subset of X and B is almost nowhere locally compact subset such that A and B are disjoint, then $D(A,B) > 0$. We will study more properties of these sets in Chapter 4.

Definitions 2.3. Let (X,d) be a metric space and A be a non-empty closed subset of X. Then A is called *almost nowhere locally compact* if for every $\varepsilon > 0$, the set $\{a \in A : \nu(a) \geq \varepsilon\}$ is compact.

Let B and C be disjoint subsets of (X,d). Then B and C are said to be *asymptotic* if $\forall \, \varepsilon > 0$, $\exists \, b \in B$, $c \in C$ with $d(b,c) < \varepsilon$.

Let us denote the set of almost nowhere locally compact sets of a metric space (X,d) by $AC(X)$.

Example 2.1. Consider $X = \{\frac{1}{n} : n \in \mathbb{N}\}$ with the usual distance metric. Let B be any subset of X. Note that for every $\varepsilon > 0$, the set $\{b \in B \,:\, \nu(b) \geqslant \varepsilon\}$ is finite and thus compact. So every non-empty closed subset of X belongs to $AC(X)$. On the other hand, if we consider $\mathbb{N}$ as a subset of $\mathbb{R}$ with the usual distance metric, then $\mathbb{N}$ is closed in $\mathbb{R}$ but $\mathbb{N} \notin AC(X)$ as $\nu(n) = \infty$ for all $n \in \mathbb{N}$.

Theorem 2.7. *Let (X,d) be a metric space. Then the following statements are equivalent:*

- (a) *The metric space (X,d) is cofinally complete.*
- (b) *Whenever (x_n) is a sequence in X with $\lim_{n\to\infty} \nu(x_n) = 0$, then (x_n) has a cluster point.*
- (c) *Either X is uniformly locally compact or else $nlc(X)$ is nonempty and compact and $\langle\{x : \nu(x) \leq \frac{1}{n}\}\rangle$ converges to $nlc(X)$ in Hausdorff distance.*
- (d) *The metric space (X,d) is complete and for every $\varepsilon > 0$ there exists $\delta > 0$ such that $\alpha(A) < \varepsilon$ whenever $\overline{\nu}(A) < \delta$ for any subset A of X.*
- (e) *The set $nlc(X)$ is compact and $\forall\, \varepsilon > 0$, $(nlc(X)^{\varepsilon})^c$ is uniformly locally compact in its relative topology.*
- (f) *Whenever F_1 and F_2 are disjoint asymptotic closed subsets of X, there exists $\delta > 0$ such that $F_1 \cap \{x \in X : \nu(x) > \delta\}$ and $F_2 \cap \{x \in X : \nu(x) > \delta\}$ are asymptotic.*
- (g) *Whenever B and C are non-empty disjoint closed subsets of X such that C is almost nowhere locally compact, then $D(B,C) > 0$.*
- (h) *Whenever $\langle F_n\rangle$ is a decreasing sequence of non-empty closed subsets of X with $\lim_{n\to\infty} \underline{\nu}(F_n) = 0$, then $\bigcap\{F_n : n \in \mathbb{N}\}$ is non-empty.*
- (i) *Whenever $\langle F_n\rangle$ is a decreasing sequence of non-empty closed subsets of X with $\lim_{n\to\infty} \overline{\nu}(F_n) = 0$, then $\bigcap\{F_n : n \in \mathbb{N}\}$ is non-empty.*

Proof. $(a) \Rightarrow (b)$: Suppose (x_n) is a sequence in X with $\lim_{n\to\infty} \nu(x_n) = 0$ that has no cluster point. By passing to a subsequence, we can assume that $\nu(x_n) < \frac{1}{n}$ and there exists a sequence $(w_j^n)_{j\in\mathbb{N}}$ in $B(x_n, \frac{1}{n})$ with no cluster point. Now consider a partition $\{M_n : n \in \mathbb{N}\}$ of $\mathbb{N}$ where each M_n is an infinite subset of $\mathbb{N}$. Define a sequence (y_j) where $y_j = w_j^n$ for $j \in M_n$. Then (y_j) is a cofinally Cauchy sequence in (X,d) which does not cluster, a contradiction.

$(b) \Rightarrow (c)$: Suppose X is not uniformly locally compact, then the closed set $F_n = \{x : \nu(x) \leq \frac{1}{n}\}$ is non-empty for each n. Let $x_n \in F_n$ for $n \in \mathbb{N}$. Then by (b), (x_n) has a cluster point which must lie in $nlc(X)$. Thus, $nlc(X)$ is non-empty and clearly it is compact by (b). Now for proving that the sequence $\langle F_n \rangle$ converges to $nlc(X) = \bigcap_{n\in\mathbb{N}} F_n$ in Hausdorff distance, it is enough to prove that for each $\varepsilon > 0$ there exists $n \in \mathbb{N}$ such that $F_n \subseteq (nlc(X))^{\varepsilon}$. If possible, suppose there exists $\varepsilon_o > 0$ such that for every $n \in \mathbb{N}$, there exists $a_n \in F_n$ with $d(a_n, nlc(X)) \geq \varepsilon_o$. Then by (b), (a_n) has a cluster point which lies in $nlc(X)$. Thus we get a contradiction.

$(c) \Rightarrow (d)$: If X is uniformly locally compact, then evidently (d) holds. Now suppose $nlc(X)$ is nonempty and compact and $\langle \{x : \nu(x) \leq \frac{1}{n}\} \rangle$ converges to $nlc(X)$ in Hausdorff distance. First we prove that (X,d) is complete. Let (x_n) be a Cauchy sequence in (X,d) which does not cluster. Then $\lim_{n\to\infty} \nu(x_n) = 0$ and hence by passing to a subsequence, we can assume that $\nu(x_n) \leq \frac{1}{n}$. This implies that $x_n \in F_n$, where $F_n = \{x : \nu(x) \leq \frac{1}{n}\}$. Consequently by (c), there exists a subsequence $(x_{n_k})_{k\in\mathbb{N}}$ of (x_n) such that $d(x_{n_k}, y_{n_k}) < \frac{1}{k}$ for all $k \in \mathbb{N}$, where $(y_{n_k})_{k\in\mathbb{N}}$ is a sequence in $nlc(X)$. By the compactness of $nlc(X)$, the sequence $(y_{n_k})_{k\in\mathbb{N}}$ clusters and hence (x_n) has a cluster point which is a contradiction. Thus, (X,d) is complete.

Now let $\varepsilon > 0$. Then by (c), $\exists\ \delta > 0$ such that $\{x : \nu(x) < \delta\} \subseteq nlc(X)^{\varepsilon/3}$. Since $nlc(X)$ is totally bounded, $\alpha(nlc(X)^{\varepsilon/3}) < \varepsilon$ and hence $\alpha(A) < \varepsilon\ \forall A \subseteq X$ with $\overline{\nu}(A) < \delta$.

$(d) \Rightarrow (e)$: Being a closed subset of a complete space, the set $nlc(X)$ is complete. Moreover by (d), $\alpha(nlc(X)) = 0$ as $\overline{\nu}(nlc(X)) = 0$ and hence $nlc(X)$ is totally bounded as well. For the other part, suppose $\exists\ \varepsilon_o > 0$ such that $(nlc(X)^{\varepsilon_o})^c$ is not uniformly locally compact in its relative topology. Then there exists a sequence (x_n) in $A = (nlc(X)^{\varepsilon_o})^c$ such that $0 < \nu_A(x_n) < \frac{1}{n}$, where $\nu_A(.)$ is the local compactness functional in A. Since A is closed in X, $\nu(x) \leq \nu_A(x)$ for all $x \in A$ and hence $\nu(x_n) < \frac{1}{n}$. By (d), for every $m \in \mathbb{N}$, there exists $n_m \in \mathbb{N}$ such that $n_m > n_{m-1}$ $(n_o = 1)$ and $\alpha(A_{n_m}) < \frac{1}{m}$, where $A_{n_m} = \{x_n : n \geq n_m\}$. Thus there exists an infinite subset $A^1_{n_1}$ of A_{n_1} which is contained in a ball of radius 1. Note that $\exists$ a finite subset F of $A^1_{n_1}$ such that $A^1_{n_1} \setminus F \subseteq A_{n_2}$. Hence there exists an infinite subset $A^2_{n_2}$ of $A^1_{n_1}$ which is contained in a ball of radius $1/2$. Proceeding by induction, we will get a decreasing sequence $(A^m_{n_m})$ of infinite subsets of X such that $A^m_{n_m}$ is contained in a ball of radius $1/m$. Now pick $y_m \in A^m_{n_m}$ for each $m \in \mathbb{N}$. Then (y_m) is Cauchy and hence it clusters, which implies that the sequence (x_n) has a cluster point, say x. Since $\nu(x_n) < \frac{1}{n}$, $x \in nlc(X)$. We arrive at a contradiction since (x_n) is a sequence in the closed set $(nlc(X)^{\varepsilon_o})^c$.

$(e) \Rightarrow (a)$: If $nlc(X) = \emptyset$, then (e) means that X is uniformly locally compact and hence (X,d) is cofinally complete. Otherwise $nlc(X)$ is nonempty and compact. Let (x_n) be a cofinally Cauchy sequence without a constant subsequence, and let $\{\mathbb{M}_j : j \in \mathbb{N}\}$ be the family of infinite subsets of $\mathbb{N}$ for (x_n)

described in Proposition 1.1. If for some $\varepsilon > 0$ and for infinitely many $j \in \mathbb{N}$ $\{x_n : n \in \mathbb{M}_j\} \cap (nlc(X)^{\varepsilon})^c$ is infinite, then by (e), (x_n) has a cluster point by (relative) uniform local compactness of $(nlc(X)^{\varepsilon})^c$. Otherwise, for each $\varepsilon > 0, \exists\ j_o \in \mathbb{N}$ such that $j \geq j_o$ implies $\{x_n : n \in \mathbb{M}_j\} \cap (nlc(X)^{\varepsilon})^c$ is a finite set. Thus, $\forall\ \varepsilon > 0$, $(nlc(X))^{\varepsilon}$ contains infinitely many terms of (x_n), and by compactness (x_n) has a cluster point in $nlc(X)$. This proves that (X,d) is cofinally complete.

$(b) \Rightarrow (f)$: Suppose no such δ exists. Then for each $n \in \mathbb{N}$, there exists $x_n \in F_1$ and $w_n \in F_2$ with $d(x_n, w_n) < \frac{1}{n}$ and $\{x_n,\ w_n\} \cap \{x : \nu(x) \leq \frac{1}{n}\} \neq \emptyset$. By the uniform continuity of the functional ν, $\lim_{n\to\infty} \nu(x_n) = \lim_{n\to\infty} \nu(w_n) = 0$. Consequently, by (b) (x_n) and (w_n) has a cluster point which will lie in $F_1 \cap F_2$. We arrive at a contradiction.

$(f) \Rightarrow (g)$: Suppose $D(B,C) = 0$. Then by (f), there exists $\delta > 0$ such that for all $n \in \mathbb{N}$, $\exists\ b_n \in B,\ c_n \in C$ with $\nu(b_n) > \delta$, $\nu(c_n) > \delta$ and $d(b_n, c_n) < \frac{1}{n}$. Since C is almost nowhere locally compact, the sequence (c_n) and hence (b_n) has a cluster point which lies in $B \cap C$. We arrive at a contradiction.

$(g) \Rightarrow (h)$: By passing to a subsequence, we can assume that $\underline{\nu}(F_n) < \frac{1}{n}$ and hence for each $n \in \mathbb{N}$, there exists $x_n \in F_n$ with $\nu(x_n) < \frac{1}{n}$. If (x_n) has a cluster point then it must lie in $\bigcap\{F_n : n \in \mathbb{N}\}$, as $\langle F_n \rangle$ is a decreasing sequence of closed sets. So assume that (x_n) has no cluster point. Without loss of generality, we can assume that (x_n) has all distinct terms. Let $C = \{x_n : n \in \mathbb{N}\}$. Then C is an almost nowhere locally compact set as for each $\varepsilon > 0$, the set $\{x \in (x_n) : \nu(x) \geq \varepsilon\}$ is finite. If C is totally bounded, then we can find a Cauchy sequence (c_n) in C with distinct terms. Thus, the set $B_1 = \{c_{2n} : n \in \mathbb{N}\}$ is a closed subset of X which is disjoint from the almost nowhere locally compact set $C_1 = \{c_{2n+1} : n \in \mathbb{N}\}$ but $D(B_1, C_1) = 0$. We get a contradiction. Thus C is not totally bounded and hence we can find a sequence (a_n) in C and $\delta > 0$ such that the family of balls $\{B(a_n, \delta) : n \in \mathbb{N}\}$ is pairwise disjoint. Since $\lim_{n\to\infty} \nu(a_n) = 0$, by passing to a tail of the sequence we can assume that for each n, $\exists\ e_n \in B(a_n, \delta) \setminus \{a_n\}$ with $\lim_{n\to\infty} d(a_n, e_n) = 0$. Consequently, $\lim_{n\to\infty} \nu(e_n) = 0$ and hence the set $E = \{e_n : n \in \mathbb{N}\}$ is almost nowhere locally compact. But $D(A,E) = 0$, where $A = \{a_n : n \in \mathbb{N}\}$, a contradiction to (g).

$(h) \Rightarrow (i)$: This is immediate.

$(i) \Rightarrow (b)$: Let (x_n) be a sequence in X with $\lim_{n\to\infty} \nu(x_n) = 0$. Without loss of generality, we may assume that for each n, $\nu(x_n) \geq \nu(x_{n+1})$. For each n, set $F_n = \overline{\{x_k : k \geq n\}}$. Then $\overline{\nu}(F_n) = \overline{\nu}(\{x_k : k \geq n\}) = \nu(x_n)$ and hence by (i) the set of cluster points of (x_n) which is precisely $\bigcap\{F_n : n \in \mathbb{N}\}$ is non-empty. Thus (x_n) has a cluster point as required. □

Remark 2.4. Theorem 2.7 is due to [Beer (2008)] except for the equivalence of (g) and (a), which was proved in [Beer and Di Maio (2012)].

As a consequence of the previous result, the first thing one should look for to spot a space that is not cofinally complete is non-compactness of $nlc(X)$. Moreover, recall that if $(X, ||\cdot||)$ is an infinite dimensional normed linear space then each closed ball in X is non-compact. Thus, $nlc(X) = X$ which is not compact and hence X is not cofinally complete. On the other hand, every finite dimensional normed linear space is uniformly locally compact and hence it is cofinally complete.

Note that by dropping the completeness in condition (d) of Theorem 2.7, we need not get a characterization of a metric space whose completion is cofinally complete. For example, consider the set of rationals, $\mathbb{Q}$ with the usual distance metric. Since $\nu(a) = 0\,\forall\, a \in \mathbb{Q}$, $\overline{\nu}(A) = 0\,\forall\, A \subseteq \mathbb{Q}$. Now consider any unbounded subset of $\mathbb{Q}$, say B, then certainly $\alpha(B) > 0$. Thus, for $\varepsilon_o = \alpha(B) > 0$, there does not exist any $\delta > 0$ such that $\overline{\nu}(A) < \delta$ implies $\alpha(A) < \varepsilon_o$ for $A \subseteq \mathbb{Q}$.

Evidently, Theorem 2.7 shows the significance of the local compactness functional in characterizing cofinally complete metric spaces. Now, in the next section, we mainly focus on equivalent conditions for a metric space to be cofinally complete in terms of certain functions.

2.3 Cofinal Completeness vis-à-vis Functions between Metric Spaces

In this section, we study various nice properties that every continuous function (or its various thin subclasses) enjoys on a cofinally complete metric space. Let us now state some definitions that are required for our next set of characterizations of cofinally complete metric spaces.

The following is a very interesting property of continuous functions that characterizes cofinally complete metric spaces, given in [Beer (2008)].

Definition 2.4. A function $g : (X,d) \to (Y,\rho)$ between two metric spaces is called *uniformly locally bounded* if $\exists\, \delta > 0$ such that $\forall\, x \in X$, $g(B_d(x,\delta))$ is a bounded subset of (Y,ρ).

A few years later, in [Aggarwal and Kundu (2016)], CC-regular functions and characterized cofinally complete metric spaces were defined in a similar way. Recall that a CC-regular function between two metric spaces need not be even continuous. For example, let $X = \{0, \frac{1}{n+1} : n \in \mathbb{N}\}$ and $Y = \{\frac{1}{n} : n \in \mathbb{N}\}$ with the usual metric and let $f : X \to Y$ be defined as:

$$f(x) = \begin{cases} 1 : x = 0 \\ x : \text{otherwise} \end{cases}$$

It is evident that every continuous function on a cofinally complete space with values in a metric space is CC-regular. The next theorem says that the converse is also true. In the same theorem, it is also proved that verifying the CC-regularity of a smaller class of continuous functions, the class of locally Lipschitz functions, is sufficient for proving cofinal completeness of a metric space. In the next chapter, we will study the conditions under which every Cauchy-regular function is CC-regular.

In [Beer and Garrido (2014)], the authors showed that every real-valued Cauchy-regular function on a metric space (X,d) is bounded if and only if (X,d) is totally bounded. We know that each real-valued continuous function defined on a metric space (X,d) is bounded if and only if (X,d) is compact. Also, a metric space (X,d) is boundedly compact if and only if each continuous function defined on it is bounded on bounded subsets of X. Our next theorem also studies the boundedness of real-valued continuous functions, defined on cofinally metric spaces, on some particular type of sets.

Definition 2.5. Let (X,d) be a metric space and let A be a non-empty subset of (X,d). Then a real-valued function f on X is said to be ν-*bounded* on A if for some $r > 0$, the set $\{f(x) : x \in A,\ \nu(x) < r\}$ is bounded. We call f to be ν-bounded if it is ν-bounded on X.

Note that a ν-bounded function need not be bounded in general. For example, consider any unbounded real-valued function defined on $\mathbb{R}$. Then clearly it is ν-bounded on every subset of $\mathbb{R}$ as $\nu(x) = \infty\ \forall\, x \in \mathbb{R}$. Now we are ready to study the ν-boundedness of continuous functions and prove some useful characterizations of cofinally complete metric spaces.

Theorem 2.8. *Let (X,d) be a metric space. Then the following statements are equivalent:*

(a) *The metric space (X,d) is cofinally complete.*
(b) *For each metric σ equivalent to d and each $\varepsilon > 0$, there exists $\delta > 0$ such that $\forall\, x \in X,\ B_d(x,\delta) \subseteq \bigcup_{i=1}^{n} B_\sigma(x_i,\varepsilon)$ for some finite subset $\{x_1,\ldots,x_n\}$ of X.*
(c) *Every d-cofinally Cauchy sequence (that is cofinally Cauchy with respect to metric d) in X is σ-cofinally Cauchy for all equivalent metrics σ on X.*

(d) *Each continuous function on* (X,d) *with values in a metric space* (Y,ρ) *is CC-regular.*

(e) *Each locally Lipschitz function on* (X,d) *with values in a metric space* (Y,ρ) *is CC-regular.*

(f) *Every closed and locally compact subset of* (X,d) *is uniformly locally compact in its relative topology.*

(g) *The set* $nlc(X)$ *is compact and for any sequence* (x_n) *in* $X \setminus nlc(X)$ *with no cluster point,* $\inf_n \nu(x_n) > 0$.

(h) *Let* (x_n) *be a sequence in X with no cluster point and* $B = \{n \in \mathbb{N} : x_n \in nlc(X)\}$. *Then B is finite and* $\inf\{\nu(x_n) : n \in \mathbb{N} \setminus B\} > 0$.

(i) *Let A be a subset of X with no limit point in X. Then* $A \cap nlc(X)$ *is finite and* $\inf\{\nu(x) : x \in A \setminus nlc(X)\} > 0$.

(j) *Let* $f : X \to \mathbb{R}$ *be a continuous function. Then there exists* $n_o \in \mathbb{N}$ *such that* $A \cap nlc(X) = \emptyset$, *where* $A = \{x \in X : |f(x)| \geq n_o\}$, *and* $\inf\{\nu(x) : x \in A\} > 0$.

(k) *Every real-valued continuous function on* (X,d) *is* ν*-bounded on every subset of X.*

(l) *Each continuous function on* (X,d) *with values in a metric space* (Y,ρ) *is uniformly locally bounded.*

(m) *For each equivalent metric* σ, *there exists* $\delta > 0$ *such that for all* $x \in X$, $B_d(x,\delta)$ *is a* σ*-bounded set.*

(n) *Each locally Lipschitz function on* (X,d) *with values in a metric space* (Y,ρ) *is uniformly locally bounded.*

(o) *Each locally Lipschitz function on* (X,d) *with values in a metric space* (Y,ρ) *is uniformly locally Lipschitz.*

Proof. The implications $(d) \Rightarrow (e)$, $(g) \Rightarrow (h) \Rightarrow (i)$ and $(o) \Rightarrow (n)$ are immediate.

$(a) \Rightarrow (b)$: Since a cofinally complete metric space is uniformly paracompact, the result immediately follows by applying Corollary 2.1 to the open cover $\{B_\sigma(x,\varepsilon) : x \in X\}$ of X.

$(b) \Rightarrow (c)$: Let σ be a metric on X equivalent to d and let (x_n) be d-cofinally Cauchy in X. If $\varepsilon > 0$, then there exists a $\delta > 0$ such that $\forall\, x \in X,\ B_d(x,\delta) \subseteq \bigcup_{i=1}^n B_\sigma(x_i,\varepsilon)$ for some finitely many x_i's in X. For this $\delta > 0$, we have some $p \in X$ and an infinite subset N of $\mathbb{N}$ such that $x_n \in B_d(p,\delta)\ \ \forall\, n \in N$. Consequently, x_n belongs to a σ-ball of radius ε for infinitely many n and hence (x_n) is σ-cofinally Cauchy.

$(c) \Rightarrow (d)$: Let $f : (X,d) \to (Y,\rho)$ be a continuous function. Now, define a metric σ on X as follows:

$$\sigma(a,b) = d(a,b) + \rho(f(a),f(b)) \text{ for } a,\ b \text{ in } X$$

Then σ is equivalent to d. Let (x_n) be a d-cofinally Cauchy sequence in X, then it is σ-cofinally Cauchy. Hence by the definition of σ, $(f(x_n))$ is cofinally Cauchy in (Y,ρ).

$(e) \Rightarrow (f)$: Let A be a closed and locally compact subset of X such that $\inf\{\nu_A(x) : x \in A\} = 0$, where $\nu_A(.)$ denotes the local compactness functional on (A,d). Then for every $n \in \mathbb{N}$, $\exists\, x_n \in A$ such that $\nu_A(x_n) < \frac{1}{n}$, which implies that $\lim_{n\to\infty} \nu_A(x_n) = 0$. If possible, suppose that (x_n) does not have a cluster point in A. Since $\nu_A(x_n) < \frac{1}{n}$, let $(a_m^n)_{m\in\mathbb{N}}$ be a sequence of distinct points in $C(x_n, \frac{1}{n}) \cap A$ with no cluster point. Then $B = \{x_n, a_m^n : m,\ n \in \mathbb{N}\}$ is a closed discrete subset of X. Since B is countable, let $\{y_n : n \in \mathbb{N}\}$ be an enumeration of B. Consider the function $f : B \to \mathbb{R}$: $f(y_n) = n$. Then f is a continuous function on a closed set B and hence by Tietze's extension theorem, f can be extended to a real-valued continuous function f' on X. By Theorem 1.4, for $0 < \varepsilon < 1$ there exists a locally Lipschitz function $g : (X,d) \to \mathbb{R}$ with $\sup_{x\in X} |f'(x) - g(x)| \leq \frac{\varepsilon}{3}$. Thus by (e), g is CC-regular. Since (y_n) is cofinally Cauchy, $(g(y_n))$ is also cofinally Cauchy. Consequently, there exists an infinite subset N of $\mathbb{N}$ such that $|g(y_n) - g(y_m)| < \frac{\varepsilon}{3}$ $\forall\, n,\ m \in N$. Hence

$$
\begin{aligned}
1 \leq |n-m| &= |f(y_n) - f(y_m)| \\
&\leq |f(y_n) - g(y_n)| + |g(y_n) - g(y_m)| + |g(y_m) - f(y_m)| \\
&< \varepsilon < 1 \quad \forall\, n,\ m \in N,\ n \neq m
\end{aligned}
$$

We arrive at a contradiction. Thus, the sequence (x_n) has a cluster point, say x, which implies that there exists a subsequence (x_{n_k}) of (x_n) converging to x. Since ν_A is a continuous function, we get $\nu_A(x) = 0$. But this contradicts the fact that A is locally compact. Consequently, $\inf\{\nu_A(x) : x \in A\} > 0$.

$(f) \Rightarrow (g)$: Let (x_n) be a sequence of distinct points in $nlc(X)$ having no cluster point. Thus, $\forall\ k \in \mathbb{N}$, $C(x_k, \frac{1}{k})$ is not compact, that is, there exists a sequence $(y_m^k)_{m\in\mathbb{N}} \subseteq C(x_k, \frac{1}{k})$ of distinct terms with no cluster point. Let $A = \{x_k,\ y_m^k : k,\ m \in \mathbb{N}\}$. Then A is closed and locally compact in its relative topology because each x_i has a neighbourhood, say $B(x_i, r_i)$, that does not contain any other element of the sequence (x_n) and $C(x_n, \frac{1}{n}) \cap B(x_i, \frac{r_i}{2}) \neq \emptyset$ for at most finitely many n. Hence, there exists a neighbourhood of x_i which contains no other element of A. Similarly, we can prove that $\nu_A(y_m^k) > 0\ \ \forall\ k,\ m \in \mathbb{N}$. Thus, A is uniformly locally compact in its relative topology. Let $\nu_A(x) > \frac{1}{n_o}\ \forall\, x \in A$, for some $n_o \in \mathbb{N}$. Then $C(x_k, \frac{1}{k}) \cap A$ is compact $\forall\, k \geq n_o$, which is a contradiction. Thus, (x_n) has a cluster point and hence $nlc(X)$ is compact.

Now let (a_n) be a sequence in $X \setminus nlc(X)$ with no cluster point. We claim that $\inf_n \nu(a_n) > 0$. Suppose that for all $k \in \mathbb{N}$, $\exists\, n_k \in \mathbb{N}$ with $0 < \nu(a_{n_k}) < \frac{1}{k}$. Then, $C(a_{n_k}, \frac{1}{k})$ is not compact. Now the arguments, similar to those we used for showing $nlc(X)$ compact, will give us a contradiction.

$(i) \Rightarrow (j)$: Suppose that $\forall\, n \in \mathbb{N}$, $\exists\, x_n \in X$ such that $|f(x_n)| \geq n$ and $\nu(x_n) = 0$. Then the sequence (x_n) does not have a cluster point, because f is continuous and every subsequence of $(f(x_n))$ is unbounded. Hence the set $B = \{x_n : n \in \mathbb{N}\}$ does not have a limit point. But $\nu(x_n) = 0\ \forall\, n \in \mathbb{N}$ which is a contradiction to (i). Thus, there exists $n_o \in \mathbb{N}$ such that every point of $A = \{x : |f(x)| \geq n_o\}$ does not belong to $nlc(X)$. Now we claim that $\inf\{\nu(x) : x \in A\} > 0$. Let $\inf\{\nu(x) : x \in A\} = 0$. Thus, $\forall\, n \in \mathbb{N}$, $\exists\, a_n \in A$ such that $\nu(a_n) < \frac{1}{n}$. Then the sequence (a_n) does not have a cluster point as A is closed and ν is a continuous function. Let $A_1 = \{a_n : n \in \mathbb{N}\}$. Then $\inf\{\nu(x) : x \in A_1 \setminus nlc(X)\} = 0$. But this gives a contradiction to (i). Hence, $\inf\{\nu(x) : x \in A\} > 0$.

$(j) \Rightarrow (k)$: Let $B \subseteq X$. By (j), we get $n_o \in \mathbb{N}$ such that $A \cap nlc(X) = \emptyset$ and $\inf\{\nu(x) : x \in A\} > 0$, where $A = \{x \in X : |f(x)| \geq n_o\}$. Let $\inf\{\nu(x) : x \in A\} = r > 0$. Then $\{f(x) : x \in B,\ \nu(x) < r\}$ is bounded as $\nu(x) < r \Rightarrow x \in X \setminus A$.

$(k) \Rightarrow (l)$: Let $f : (X,d) \to (Y,\rho)$ be a continuous function which fails to be uniformly locally bounded. Then there exists a sequence (x_n) in X such that $f(B_d(x_n, \frac{1}{n}))$ is an unbounded subset of (Y,ρ). Thus $C_d(x_n, \frac{1}{n})$ is not compact and hence $\nu(x_n) \leq \frac{1}{n}$. Similarly, since $B_d(x_n, \frac{1}{n}) \subseteq B_d(x, \frac{2}{n})$ for every $x \in B_d(x_n, \frac{1}{n})$, $\nu(x) \leq \frac{2}{n}$. Now define the continuous function $g : (Y,\rho) \to \mathbb{R}$ by $g(y) = \rho(y,a)$, where a is some arbitrary point of Y. Since $g \circ f$ is a real-valued continuous function on (X,d), by (k) it is ν-bounded on X. Subsequently, there exists $r > 0$ such that the set $\{g \circ f(x) : x \in X,\ \nu(x) < r\}$ is a bounded subset of $\mathbb{R}$. So there exist $M > 0$ and $n_o \in \mathbb{N}$ such that $g \circ f(x) \leq M\ \ \forall\, x \in B_d(x_n, \frac{1}{n})$ and $\forall\, n \geq n_o$. This implies that $\rho(f(x),a) \leq M\ \ \forall\, x \in B_d(x_n, \frac{1}{n})$ and $\forall\, n \geq n_o$. Thus we get a contradiction to the choice of (x_n). Consequently, f is uniformly locally bounded.

$(l) \Rightarrow (m)$: Consider the identity map, $id : (X,d) \to (X,\sigma)$.

$(m) \Rightarrow (n)$: Similar to the proof of the implication $(c) \Rightarrow (d)$.

$(n) \Rightarrow (a)$: If (X,d) is not cofinally complete, then there exists a cofinally Cauchy sequence (x_n) of distinct points in (X,d) with no cluster point. Then the set $A = \{x_n : n \in \mathbb{N}\}$ is closed and discrete and hence the function $f : (A,d) \to \mathbb{R}$ defined by $f(x_n) = n$ is continuous. By Tietze's extension theorem, there exists a real-valued continuous extension f' of f to (X,d). Then Theorem 1.4 implies that for $\varepsilon > 0$, there exists a locally Lipschitz function $g : (X,d) \to \mathbb{R}$ with $\sup_{x \in X} |f'(x) - g(x)| \leq \varepsilon$. By (n), g is uniformly locally bounded and hence $\exists\, \delta > 0$ such that $\forall\, x \in X$, $g(B(x,\delta))$ is a bounded subset of $\mathbb{R}$. Since (x_n) is cofinally Cauchy, there exists an infinite subset N of $\mathbb{N}$ such that for every $m \in N$, $x_m \in B(x_{n_o}, \delta)$ for some $n_o \in \mathbb{N}$. Since $m = f'(x_m) \leq g(x_m) + \varepsilon < M + \varepsilon$, for every $m \in N$ and for some $M > 0$, we get a contradiction. Thus (X,d) is cofinally complete.

$(a) \Rightarrow (o)$: Let $f : (X,d) \to (Y,\rho)$ be locally Lipschitz. Fix $y_o \in Y$; for each $n \in \mathbb{N}$, let V_n be the following open subset of X:

$$V_n := \left\{x : \exists\, \varepsilon > 0\, \forall\, a,\, b \in B_d\left(x, \frac{1}{n} + \varepsilon\right),\, \rho(f(a), f(b)) < nd(a,b),\, f(x) \in B_\rho(y_o, n)\right\}$$

Clearly, $\{V_n : n \in \mathbb{N}\}$ is an open cover of X directed by inclusion. Since a cofinally complete metric space is uniformly paracompact, by Corollary 2.1 there exists $\mu > 0$ such that each open ball of radius μ has a finite subcover from $\{V_n : n \in \mathbb{N}\}$, and since the cover is directed by inclusion, each such ball is actually contained in some single member of the cover. Fix $x \in X$ and choose $k \in \mathbb{N}$ such that $B_d(x,\mu) \subseteq V_k$.

Now we claim that f is Lipschitz on $B_d(x,\mu)$. Suppose $\{a,b\} \subseteq B_d(x,\mu)$. If $d(a,b) < \frac{1}{k}$, then by definition of V_k, we have $\rho(f(a), f(b)) < kd(a,b)$. On the other hand, if $d(a,b) \geq \frac{1}{k}$, then $\{f(a), f(b)\} \subseteq B_\rho(y_o, k)$. So we have

$$\frac{\rho(f(a), f(b))}{d(a,b)} < \frac{2k}{1/k} = 2k^2$$

Thus f restricted to $B_d(x,\mu)$ is Lipschitz with Lipschitz constant $2k^2$. □

Remark 2.5. (i) The conditions (d), (f) and (h) to (k) were proved to be equivalent to (a) in [Aggarwal and Kundu (2016)]. While the conditions (l), (m) and (o) in Theorem 2.8 were proved to be equivalent to (a) in [Beer (2008)], [Beer and Di Maio (2012)] and [Beer and Garrido (2015)] respectively.
(ii) Theorem 2.8 still holds if we replace the metric space (Y,ρ) by $(\mathbb{R}, |\cdot|)$.

It remained unnoticed that the class of CC-regular functions is actually contained in the class of uniformly locally bounded functions. The following result is given in [Gupta and Kundu (2020)].

Theorem 2.9. *[Gupta and Kundu (2020)] If f is a CC-regular function from a metric space (X,d) to another metric space (Y,ρ), then f is uniformly locally bounded.*

Proof. Let $f : (X,d) \to (Y,\rho)$ be a CC-regular function. Suppose f is not uniformly locally bounded. Therefore $\forall n \in \mathbb{N}$, $\exists x_n \in X$ such that $f(B(x_n, \frac{1}{n}))$ is not a bounded subset of Y. Let ε be any positive real number. Now, $\forall n \in \mathbb{N}$, $f(B(x_n, \frac{1}{n}))$ is not bounded in Y. Therefore, $\forall n \in \mathbb{N}$, there exits a sequence

$(f(x_n^m))_{m\in\mathbb{N}}$ in $f\left(B\left(x_n, \frac{1}{n}\right)\right)$ such that $\rho\left(f(x_n^m), f(x_n^t)\right) \geq \varepsilon$ for all $m, t \in \mathbb{N}$. Let $A_n = \{f(x_n^m) : m \in \mathbb{N}\}$. Now we will construct a sequence in Y by induction. Let $F_1 = \{f(x_1^1)\}$. If $d(f(x_1^1), y) < \frac{\varepsilon}{2}$ for every $y \in A_2$, then $d(y, y') < \varepsilon$ for all $y, y' \in A_2$. We get a contradiction. Let $f(x_2^2)$ (rename if necessary) $\in A_2$ such that $\rho(f(x_1^1), f(x_2^2)) \geq \frac{\varepsilon}{2}$. Similarly, we can choose $f(x_1^2) \in A_1 \setminus \{f(x_1^1), f(x_2^2)\}$ such that $F_2 = \{f(x_1^1), f(x_2^2), f(x_1^2)\}$ is $\frac{\varepsilon}{2}$ discrete. Now suppose a finite subset F_n of $\bigcup_{i=1}^{n} A_i$ is chosen satisfying: 1) F_n is $\frac{\varepsilon}{2}$ discrete, 2) $|F_n \cap A_i| = n - i + 1$ for all $1 \leq i \leq n$. To construct F_{n+1}, choose $f(x_{n+1}^{n+1}) \in A_{n+1}$ such that $F_n \cup \{f(x_{n+1}^{n+1})\}$ is $\frac{\varepsilon}{2}$ discrete. Suppose it is not possible, then $\rho(y, F_n) < \frac{\varepsilon}{2}$ for all $y \in A_{n+1}$. Since A_{n+1} is infinite and F_n is finite, there exists $y,\ y' \in A_{n+1} \setminus F_n$, $y \neq y'$ such that $\rho(y, z) < \frac{\varepsilon}{2}$ and $\rho(y', z) < \frac{\varepsilon}{2}$ for some $z \in F_n$. This implies $\rho(y, y') < \varepsilon$, but it gives a contradiction. Repeating this process, we construct $F_{n+1} = F_n \bigcup \left\{f(x_1^{n+1}), f(x_2^{n+1}), \ldots, f(x_{n+1}^{n+1})\right\}$ such that F_{n+1} is $\frac{\varepsilon}{2}$ discrete. Now if we take a sequence in the order we chose the elements of the type $f(x_j^i)$, $i,\ j \in \mathbb{N}$, the sequence is $\frac{\varepsilon}{2}$ discrete but its pre-image is cofinally Cauchy as it consists of infinite elements from each ball $B(x_n, \frac{1}{n})$. Since f is CC-regular, we get a contradiction. □

Remark 2.6. A uniformly locally bounded function between two metric spaces need not be CC-regular. For example, consider $X = \{\frac{1}{n} : n \in \mathbb{N}\}$ with the usual distance metric d and $Y = \{n : n \in \mathbb{N}\}$ with the $\{0, 1\}$ discrete metric ρ. Define f from (X, d) to (Y, ρ) such that $f(\frac{1}{n}) = n$ for all $n \in \mathbb{N}$.

Note that the above example shows that even a continuous uniformly locally bounded function need not be CC-regular. But for a special kind of range, we have the following result.

Proposition 2.1. *[Gupta and Kundu (2020)] Let $f : (X, d) \to (Y, \rho)$ be a function between two metric spaces. Suppose every bounded subset of Y is totally bounded. Then f is CC-regular if and only if it is uniformly locally bounded.*

We know that a metric space is UC if each continuous function defined on it is uniformly continuous. Since cofinally complete metric spaces are weaker than UC spaces, it is natural to investigate some particular subset of the set of continuous functions on a cofinally complete metric space such that every function from that subset is uniformly continuous. To solve this purpose Beer considered the following subsets [Beer (2008)].

Definition 2.6. Let (X,d) and (Y,ρ) be metric spaces. Then $f \in CV(X,Y)$ if f is continuous and for all $\varepsilon > 0, f$ is uniformly continuous on the set $\{x \in X : \nu(x) > \varepsilon\}$.

Example 2.2. Take any continuous function g on $\mathbb{R}$ which is not uniformly continuous, clearly $g \notin CV(X,\mathbb{R})$. Consider $X = \{e_n + \frac{1}{n}e_k : n,k \in \mathbb{N}\}$ as a metric subspace of the real Hilbert space ℓ_2, where $\{e_n : n \in \mathbb{N}\}$ is the standard orthonormal basis of ℓ_2. Let $A = \{e_n + \frac{1}{n}e_1 : n \in \mathbb{N}\}$. Define a function $f : (X,d) \to \mathbb{R}$ such that $f(a) = 1$ if $a \in A$, otherwise 0. It can be verified that $f \in CV(X,\mathbb{R})$ but f is not uniformly continuous.

Theorem 2.10. *Let (X,d) be a metric space. Then the following statements are equivalent.*

- (a) *The metric space (X,d) is cofinally complete.*
- (b) *Whenever (Y,ρ) is a metric space and $f \in CV(X,Y)$, then f is uniformly continuous on (X,d).*
- (c) *Whenever (Y,ρ) is a metric space and f is a locally Lipschitz function belonging to $CV(X,Y)$, then f is uniformly continuous on (X,d).*
- (d) *If f is a locally Lipschitz function belonging to $CV(X,\mathbb{R})$, then f is uniformly continuous on (X,d).*

Remark 2.7. The equivalence of the statements (a) and (b) of the previous theorem was proved in Theorem 3.5 of [Beer (2008)]. The proof of the implication $(d) \Rightarrow (a)$ is similar to that of Theorem 3.5 of [Beer (2008)], where Beer proved the same for continuous functions (instead of locally Lipschitz functions).

We know that a function $f : (X,d) \to (Y,\rho)$ between two metric spaces (X,d) and (Y,ρ) is called uniformly continuous if for all $\varepsilon > 0$, there exists $\delta > 0$ such that for each $A \subseteq X$ with the diameter of A less than δ, the diameter of $f(A)$ is less than ε, that is, $f(A) \subseteq B(b,\varepsilon)$ for some $b \in Y$. In [Keremedis (2017)], Keremedis generalized the definition of uniformly continuous functions by replacing one open ball of radius ε in the metric space Y by a union of finitely many open balls of radius ε and thus defined another class of functions called almost bounded functions. The precise definition follows.

Definition 2.7. A function f from a metric space (X,d) to a metric space (Y,ρ) is called *almost bounded* if for all $\varepsilon > 0$, there exists $\delta > 0$ such that for every $A \subseteq X$ with $d(A) < \delta$, there exists a finite subset $B = \{b_1, b_2, \ldots, b_n\}$ of Y such that $f(A) \subseteq \bigcup_{i=1}^{n} B(b_i, \varepsilon)$.

A function with totally bounded range is almost bounded in the sense of Keremedis. Note that an almost bounded function between two metric spaces need not be continuous. For example, consider $X = \{0\} \cup \{1/n : n \in \mathbb{N}\}$ with the usual distance metric and define $g(1/n) = 1 \ \forall \ n \in \mathbb{N}$ and $g(0) = 0$. As a result, Keremedis considered those functions which were continuous as well as almost bounded and called them almost uniformly continuous. Parallel to the definition of UC spaces, he defined AUC spaces as follows.

Definitions 2.8. A continuous almost bounded function from a metric space (X,d) to a metric space (Y,ρ) is called *almost uniformly continuous.*

A metric space (X,d) is said to be an *almost uniformly continuous* space or an *AUC* space if every continuous real-valued function on (X,d) is almost uniformly continuous.

Remark 2.8. Each uniformly continuous function between any two metric spaces is almost uniformly continuous, but an almost uniformly continuous function need not be even Cauchy-regular. For example, consider $X = \{\frac{1}{n} : n \in \mathbb{N}\}$ with the usual distance metric and let $f : X \to \mathbb{R}$ be defined as:

$$f\left(\frac{1}{n}\right) = \begin{cases} 1 : n \text{ is odd} \\ 2 : n \text{ is even} \end{cases}$$

In [Gupta and Kundu (2020)], it was shown that an almost bounded function between two metric spaces is nothing but a CC-regular function and an AUC space is nothing but a cofinally complete metric space.

Theorem 2.11. *[Gupta and Kundu (2020)] Let f be a function between two metric spaces (X,d) and (Y,ρ). Then f is CC-regular if and only if f is almost bounded.*

Proof. In a manner similar to the proof of Theorem 2.9, we can prove that every CC-regular function between two metric spaces is almost bounded.

Conversely, let $f : (X,d) \to (Y,\rho)$ be an almost bounded function and (x_n) be a cofinally Cauchy sequence in X having no constant subsequence. Let $\varepsilon > 0$, then there exists a $\delta > 0$ such that for all $A \subseteq X$ with $d(A) < \delta$, there exists $\{y_1, y_2, \ldots, y_n\} \subseteq Y$ such that $f(A) \subset \bigcup_{i=1}^{n} B(y_i, \varepsilon)$. Since (x_n) is cofinally Cauchy, there exists an infinite subset $\mathbb{N}_\delta$ of $\mathbb{N}$ such that $d(x_n, x_m) < \delta$ for all $n, m \in \mathbb{N}_\delta$. Let $A = \{x_n : n \in \mathbb{N}_\delta\}$. Thus $f(A) \subset \bigcup_{i=1}^{n} B(y_i, \varepsilon)$ for some $y_i \in Y$, $1 \leq i \leq n$. Thus, there exists $i \in \mathbb{N}$ such that $B(y_i, \varepsilon)$ contains elements of the type $f(x_k)$ for infinitely many $k \in \mathbb{N}$. Hence the sequence $(f(x_n))$ is cofinally Cauchy. □

Remark 2.9. Since CC-regular functions are more familiar in the literature, from now onwards we will call almost bounded functions to be CC-regular functions.

Corollary 2.2. *[Gupta and Kundu (2020)] A metric space (X,d) is an AUC space if and only if it is cofinally complete.*

Proof. Since the class of CC-regular functions and that of almost bounded functions are same, by the definition of AUC spaces and by the fact that a metric space (X,d) is cofinally complete if and only if each real-valued continuous function defined on it is CC-regular (see Theorem 2.8), we conclude that a metric space (X,d) is an AUC space if and only if it is cofinally complete. □

Remark 2.10. It is easy to see that a CC-regular function between two metric spaces need not be bounded; consider the identity function on $\mathbb{R}$ with the usual distance metric. Also note that a bounded function need not be CC-regular. For example, consider $X = \left\{\frac{1}{n} : n \in \mathbb{N}\right\}$ with the usual distance metric d and $Y = \mathbb{N}$ with the $\{0,1\}$ discrete metric d'. Let $f : (X,d) \to (Y,d')$ be defined as:

$$f\left(\frac{1}{n}\right) = n \quad \forall\, n \in \mathbb{N}$$

It is now clear that an almost uniformly continuous function is nothing but a continuous CC-regular function, that is a function which is both continuous and CC-regular. In view of this result, in this section we would like to study some properties of continuous CC-regular functions and in the process we obtain some characterizations of cofinally complete metric spaces and totally bounded metric spaces.

It is well-known that the set of all real-valued continuous functions defined on a metric space (X,d) forms a ring with respect to the usual addition and multiplication of real-valued functions, but this is not true in case of uniformly continuous functions [Beer *et al.* (2018); Cabello Sánchez (2017)]. We have the following nice observation regarding continuous CC-regular functions.

Theorem 2.12. *Let (X,d) be a metric space and A be the set of all real-valued continuous CC-regular functions on X. Then A is a ring with respect to the usual addition and multiplication of real-valued functions.*

Proof. By Proposition 2.1, a function $f : (X,d) \to \mathbb{R}$ is CC-regular if and only if it is uniformly locally bounded. Let $f, g : (X,d) \to \mathbb{R}$ be two continuous CC-regular functions. So there exists a $\delta > 0$ such that $\forall x \in X$, $f(B(x,\delta))$ and $g(B(x,\delta))$ are bounded subsets of $\mathbb{R}$. Thus $(f+g)(B(x,\delta))$ and $(f*g)(B(x,\delta))$ are also bounded subsets of $\mathbb{R}$. Thus $f+g$ and $f*g$ are also continuous CC-regular functions.

Hence the set of all real-valued continuous CC-regular functions on X forms a ring with respect to the usual addition and multiplication of real-valued functions. □

Now we are ready to study the stability of never zero continuous CC-regular functions under reciprocation.

Theorem 2.13. *Let (X,d) be a metric space. Then the following assertions are equivalent.*

(a) (X,d) is cofinally complete.
(b) Whenever $f:(X,d)\to\mathbb{R}$ is a continuous CC-regular function such that f is never zero, then $\frac{1}{f}$ is also continuous and CC-regular.
(c) Whenever $f:(X,d)\to\mathbb{R}$ is both locally Lipschitz and CC-regular such that f is never zero, then $\frac{1}{f}$ is also locally Lipschitz and CC-regular.

Proof. The implication $(a)\Rightarrow(b)$ follows from the fact that every real-valued continuous function on a cofinally complete metric space is CC-regular and the implication $(b)\Rightarrow(c)$ is immediate.

$(c)\Rightarrow(a)$: If (X,d) is not cofinally complete, then there exists a cofinally Cauchy sequence (x_n) of distinct points in (X,d) with no cluster point. Thus the set $A=\{x_n : n\in\mathbb{N}\}$ is closed. Consequently, $\forall n\in\mathbb{N}$, $\exists\varepsilon_n>0$ such that $d(x_m,x_n)>\varepsilon_n\ \forall m\neq n$. Let $\delta_n=\min\{1/n,\varepsilon_n/3\}$. Define a function $f:(X,d)\to[0,2]$ as follows:

$$f(x)=\begin{cases}\frac{1}{n}-\frac{1}{n\delta_n}d(x,x_n) & : x\in B(x_n,\delta_n) \text{ for some } n\in\mathbb{N}\\ 0 & : \text{otherwise}\end{cases}$$

Clearly, f restricted to each ball $B(x_n,\delta_n)$ is Lipschitz. Let $x\in X$. Since x is not a cluster point of the sequence (x_n) and $\varepsilon_n\leq 1/n\ \forall n\in\mathbb{N}$, $\exists\delta_x>0$ such that $B(x,\delta_x)$ intersects at most one of the balls $B(x_n,\delta_n)$. Thus f is locally Lipschitz. Since the range of f is totally bounded, f is CC-regular as well. Now define another function $g:(X,d)\to\mathbb{R}$ as follows:

$$g(x)=f(x)+d(x,A)$$

Since the sum of two real-valued locally Lipschitz functions on a metric space is also locally Lipschitz, g is locally Lipschitz. Also, by Theorem 2.12, g is CC-regular. By the hypothesis, $\frac{1}{g}$ should also be CC-regular since g is never zero. But (x_n) is a cofinally Cauchy sequence such that $\frac{1}{g}(x_n)=n\ \forall n\in\mathbb{N}$. We get a contradiction. Hence (X,d) is cofinally complete. □

Remark 2.11. Note that if the completion $(\widehat{X},d)$ of a metric space (X,d) is cofinally complete, then the following statement need not hold: whenever

$f : (X,d) \to \mathbb{R}$ is both Cauchy-regular and CC-regular such that f is never zero, then $\frac{1}{f}$ is also Cauchy-regular and CC-regular. For example, consider the identity function f on (X,d), where $X = \{1/n : n \in \mathbb{N}\}$ and d is the usual distance metric on X. The metric space (X,d) has a cofinal completion. Also, f is Cauchy-regular, CC-regular and is never zero, but $\frac{1}{f}$ is not CC-regular.

Recently in 2020, the authors in [Beer *et al.* (2020)] have characterized cofinally complete metric spaces in terms of the stability of never zero uniformly locally Lipschitz functions under reciprocation (see Exercise 2.6). Next, we would like to explore the nature of the subsets of a cofinally complete metric space on which every continuous function is strongly uniformly continuous.

Lemma 2.2. *[Beer and Di Maio (2012)] Let (X,d) be a cofinally complete metric space and $A \in AC(X)$. Then A is compact.*

Proof. Let (a_n) be any sequence in A. If $\lim_{n\to\infty} \nu(a_n) = 0$, then by Theorem 2.7, (a_n) has a cluster point in A, else there exists $\varepsilon > 0$ such that for infinitely many $n \in \mathbb{N}$, $\nu(a_n) > \varepsilon$. Hence by the definition of the elements in $AC(X)$, (a_n) has a cluster point in A. □

Theorem 2.14. *Let (X,d) be a metric space. Then the following conditions are equivalent.*

- (a) *(X,d) is cofinally complete.*
- (b) *Whenever (Y,ρ) is a metric space and $f : (X,d) \to (Y,\rho)$ is continuous, then f is strongly uniformly continuous on each member of $AC(X)$.*
- (c) *Whenever (Y,ρ) is a metric space and $f : (X,d) \to (Y,\rho)$ is continuous, then $D_d(A,B) = 0$ implies $D_\rho(f(A), f(B)) = 0$, where $A \in AC(X)$ and B is any non-empty subset of X.*
- (d) *Whenever $f : (X,d) \to \mathbb{R}$ is continuous, then $D_d(A,B) = 0$ implies $D_{|.|}(f(A), f(B)) = 0$, where $A \in AC(X)$ and B is any non-empty subset of X.*

Proof. $(a) \Rightarrow (b)$: Let $A \in AC(X)$, by Lemma 2.2, A is compact. Thus every continuous function on X is strongly uniformly continuous on each $A \in AC(X)$.

$(b) \Rightarrow (c)$: Let $f : (X,d) \to (Y,\rho)$ be a continuous function and A and B be subsets of X such that $A \in AC(X)$ and $D_d(A,B) = 0$. To show that $D_\rho(f(A), f(B)) = 0$, let $\varepsilon > 0$. Since f is strongly uniformly continuous on A, $\exists \delta > 0$ such that if $d(x,y) < \delta$ and $\{x,y\} \cap A \neq \emptyset$, then $\rho(f(x), f(y)) < \varepsilon$. Choose $a \in A$ and $b \in B$ such that $d(a,b) < \delta$ and thus $\rho(f(a), f(b)) < \varepsilon$. Consequently, $D_\rho(f(A), f(B)) = 0$.

$(c) \Rightarrow (d)$: This is immediate.

$(d) \Rightarrow (a)$: Suppose (X,d) is not cofinally complete. By Theorem 2.7 (g), there exists $A \in AC(X)$ and a closed subset B of X such that $A \cap B = \emptyset$ but $D_d(A,B) = 0$. Thus for each $n \in \mathbb{N}$, there exists $a_n \in A$ and $b_n \in B$ such that $d(a_n, b_n) < \frac{1}{n}$. Since the sequences (x_n) and (y_n) cannot have any cluster point, the sets $U = \{a_n : n \in \mathbb{N}\}$ and $V = \{b_n : n \in \mathbb{N}\}$ are closed and U being a subset of A belongs to $AC(X)$. By Tietze's extension theorem, let $f : (X,d) \to \mathbb{R}$ be a continuous function such that $f(U) = 0$ and $f(V) = 1$. Now, f is a continuous function such that $U \in AC(X)$ and $D_d(U,V) = 0$ but $D_\rho(f(U), f(V)) \neq 0$, a contradiction. □

We will end this chapter with some more properties of CC-regular functions. We know that every continuous function on a metric space is CC-regular if and only if the metric space is cofinally complete. Our next result characterizes those metric spaces (X,d) such that every CC-regular function from (X,d) to any metric space (Y,ρ) is continuous. Such metric spaces are also equivalent to those metric spaces on which every PC-regular function is continuous.

Theorem 2.15. *[Gupta and Kundu (2020)] Let (X,d) be a metric space. Then the following statements are equivalent.*

- (a) *The metric space (X,d) is discrete.*
- (b) *Whenever (Y,ρ) is a metric space and $f : (X,d) \to (Y,\rho)$ is CC-regular, then f is continuous.*
- (c) *Whenever (Y,ρ) is a metric space and $f : (X,d) \to (Y,\rho)$ is PC-regular, then f is continuous.*

Proof. The implication $(a) \Rightarrow (b)$ is immediate and the implication $(b) \Rightarrow (c)$ follows from Proposition 1.7.

$(c) \Rightarrow (a)$: Suppose (X,d) is not discrete. Then there exists a non-isolated point x in X. Hence there exists a sequence (x_n) in X such that all the elements of the sequence are distinct, $x_n \neq x \; \forall n \in \mathbb{N}$ and (x_n) converges to $x \in X$. Define $f : (X,d) \to \mathbb{R}$ such that

$$f(z) = \begin{cases} 1 : \; if \; z = x_n \text{ for some } n \\ 0 : \text{otherwise} \end{cases}$$

Clearly, f is PC-regular but not continuous. □

Analogously we can characterize those metric spaces on which every CC-regular and PC-regular functions are Cauchy-regular and uniformly continuous.

Theorem 2.16. *[Gupta and Kundu (2020)] Let (X,d) be a metric space. Then the following statements are equivalent.*

(a) *The metric space (X,d) is complete and discrete.*
(b) *Whenever (Y,ρ) is a metric space and $f:(X,d)\to(Y,\rho)$ is CC-regular, then f is Cauchy-regular.*
(c) *Whenever (Y,ρ) is a metric space and $f:(X,d)\to(Y,\rho)$ is PC-regular, then f is Cauchy-regular.*

Proof. $(a)\Rightarrow(b)$: Let $f:(X,d)\to(Y,\rho)$ be CC-regular function. Since (X,d) is discrete, f is continuous, thus Cauchy-regular because (X,d) is complete.

$(b)\Rightarrow(c)$: This follows from Proposition 1.7.

$(c)\Rightarrow(a)$: Suppose there exists a Cauchy sequence (x_n) in (X,d) such that all its elements are distinct and it does not converge. Define $f:(X,d)\to\mathbb{R}$ such that

$$f(x)=\begin{cases}1: & if\ x=x_n \text{ for some odd } n\\ 2: & if\ x=x_n \text{ for some even } n\\ 0: & \text{otherwise}\end{cases}$$

Clearly f is PC-regular but not Cauchy-regular. Thus (X,d) is complete and by Theorem 2.15, (X,d) is discrete. □

Theorem 2.17. *[Aggarwal and Kundu (2016)] Let (X,d) be a metric space. Then the following assertions are equivalent.*

(a) *The metric space (X,d) is uniformly discrete.*
(b) *Whenever (Y,ρ) is a metric space and $f:(X,d)\to(Y,\rho)$ is CC-regular, then f is uniformly continuous.*
(c) *Whenever (Y,ρ) is a metric space and $f:(X,d)\to(Y,\rho)$ is PC-regular, then f is uniformly continuous.*

Proof. The implications $(a)\Rightarrow(b)\Rightarrow(c)$ are immediate.

$(c)\Rightarrow(a)$: Suppose (X,d) is not uniformly discrete. Thus for every $n\in\mathbb{N}$ there exist $x,\ y\in X$, $x\neq y$ such that $d(x,y)<\frac{1}{n}$. For every $n\in\mathbb{N}$, let $A_n=\{(x,y)\in X\times X: x\neq y,\ d(x,y)<\frac{1}{n}\}$. Clearly, each A_n is an infinite set and (A_n) is a decreasing sequence. We claim that $\forall\, n\in\mathbb{N}$, $\exists\, x_n,\ y_n\in X$ such that $0<d(x_n,y_n)<\frac{1}{n}$, and for $n>1$, $x_n\notin\{y_1,\dots,y_{n-1}\}$ and $y_n\notin\{x_1,\dots,x_{n-1}\}$. We prove it by induction. For $n=1$, the statement is clearly true. Suppose for $n=2,\dots,k$, we have chosen $x_n,\ y_n\in X$ such that $0<d(x_n,y_n)<\frac{1}{n}$ and $x_n\notin\{y_1,\dots,y_{n-1}\}$, $y_n\notin\{x_1,\dots,x_{n-1}\}$. Suppose the statement is not true for $n=k+1$. Thus, for every $(x,y)\in A_{k+1}$, either $x\in\{y_1,\dots,y_k\}$ or $y\in\{x_1,\dots,x_k\}$. Note that if $x\in\{y_1,\dots,y_k\}$ then

$y \in \{y_1, \ldots, y_k\}$, otherwise we choose $x_{k+1} = y$ and $y_{k+1} = x$, which is a contradiction. Similarly, if $y \in \{x_1, \ldots, x_k\}$ then $x \in \{x_1, \ldots, x_k\}$. Consequently, A_{k+1} is a finite set, a contradiction. Hence the claim is proved. Now, define the function $f : X \to Y$ as:

$$f(x) = \begin{cases} y & : x = x_n \text{ for some } n \in \mathbb{N} \\ y' & : \text{otherwise} \end{cases}$$

where y and y' are distinct elements of Y.
Then f is PC-regular but not uniformly continuous. Hence we get a contradiction. □

Next we have an interesting observation related to CC-regular functions.

Proposition 2.2. *[Aggarwal and Kundu (2016)] Let $f : (X,d) \to (Y,\rho)$ be a continuous function between two metric spaces and let A be a dense subset of X. If the restriction of f to A, $f|_A$ is CC-regular, then f is CC-regular.*

Proof. Let (x_n) be a cofinally Cauchy sequence in (X,d). By the density of A, there exists a sequence $(a^n_m)_{m\in\mathbb{N}}$ in A converging to x_n for every $n \in \mathbb{N}$. Thus, for all $n \in \mathbb{N}$, the sequence $(f(a^n_m))_{m\in\mathbb{N}}$ converges to $f(x_n)$. Consequently, $\forall\, n \in \mathbb{N}$, $\exists\, m_n \in \mathbb{N}$ such that $d(a^n_{m_n}, x_n) < \frac{1}{n}$ and $\rho(f(a^n_{m_n}), f(x_n)) < \frac{1}{n}$. Hence, the sequence $(a^n_{m_n})_{n\in\mathbb{N}}$ is cofinally Cauchy. So $(f(a^n_{m_n}))_{n\in\mathbb{N}}$ is cofinally Cauchy. Subsequently, the sequence $(f(x_n))$ is cofinally Cauchy, as required. □

We know that a CC-regular function between two metric spaces need not be continuous. The following example shows that a uniformly locally Lipschitz function between two metric spaces need not be CC-regular.

Example 2.3. Consider the real Hilbert space ℓ_2. Let $X = \bigcup_{n\in\mathbb{N}} A_n$, where $A_n = \{e_n + \frac{1}{n} e_k : k \in \mathbb{N}\}$, and $\{e_n : n \in \mathbb{N}\}$ is the standard orthonormal basis of ℓ_2. Let d be the metric on X induced by the ℓ_2 norm. Consider $Y = \{n : n \in \mathbb{N}\}$ with the $\{0,1\}$ discrete metric ρ. Let $\{\mathbb{M}_j : j \in \mathbb{N}\}$ be a pairwise disjoint family of infinite subsets of $\mathbb{N}$. Let $\mathbb{M}_n = \{x^n_1, x^n_2, x^n_3, \ldots\}$ for all $n \in \mathbb{N}$. Define f from (X,d) to (Y,ρ) such that $\forall k, n \in \mathbb{N}$

$$f\left(e_n + \frac{1}{n} e_k\right) = x^n_k$$

Now for each $z \in X$, there exists exactly one natural number n such that $B(z,1) \cap A_n \neq \emptyset$. Choose $k_z > \frac{n}{\sqrt{2}}$, thus $\rho(f(u), f(w)) \leq k_z d(u,w)$, whenever $u, w \in B(z,1)$. Hence the function is uniformly locally Lipschitz. The function is not CC-regular because if we enumerate the elements of X, we will get a cofinally Cauchy sequence but its image is certainly not cofinally Cauchy.

We know that a metric space (X,d) is compact if and only if every real-valued continuous function defined on it is bounded. In Remark 2.10 we have seen that a continuous CC-regular function need not be bounded in general. So our next result characterizes in particular those metric spaces on which every real-valued continuous CC-regular function is bounded. We also study boundedness of various combinations of Lipschitz-type functions with CC-regular functions. Note that a function $f:(X,d)\to(Y,\rho)$ between two metric spaces is called bounded if $f(X)$ is bounded in (Y,ρ).

Theorem 2.18. *Let (X,d) be a metric space. Then the following statements are equivalent.*

- (a) *The metric space (X,d) is totally bounded.*
- (b) *Whenever (Y,ρ) is a metric space and $f:(X,d)\to(Y,\rho)$ is both continuous and CC-regular, then f is bounded.*
- (c) *Whenever (Y,ρ) is a metric space and $f:(X,d)\to(Y,\rho)$ is both locally Lipschitz and CC-regular, then f is bounded.*
- (d) *Whenever (Y,ρ) is a metric space and $f:(X,d)\to(Y,\rho)$ is both Cauchy-Lipschitz and CC-regular, then f is bounded.*
- (e) *Whenever (Y,ρ) is a metric space and $f:(X,d)\to(Y,\rho)$ is both uniformly locally Lipschitz and CC-regular, then f is bounded.*
- (f) *If $f:(X,d)\to\mathbb{R}$ is uniformly locally Lipschitz, then f is bounded.*

Proof. The implications $(b)\Rightarrow(c)\Rightarrow(d)\Rightarrow(e)$ are all immediate.

$(a)\Rightarrow(b)$: Let f be a continuous CC-regular function from (X,d) to (Y,ρ). By Proposition 1.4, we infer that $f(X)$ is a totally bounded subset of (Y,ρ).

$(e)\Rightarrow(f)$: Let $f:(X,d)\to\mathbb{R}$ be a uniformly locally Lipschitz function. Then f is uniformly locally bounded and by Proposition 2.1, it is CC-regular as well. Thus the implication follows.

$(f)\Rightarrow(a)$: Suppose (X,d) is not totally bounded. Therefore, $\exists\,\delta>0$ and a sequence (x_n) in X such that $d(x_n,x_m)>\delta\ \forall n,m\in\mathbb{N}\ (n\neq m)$. Define a function $f:(X,d)\to\mathbb{R}$ as follows:

$$f(x)=\begin{cases} n-\frac{4n}{\delta}d(x,x_n) & : x\in B\left(x_n,\frac{\delta}{4}\right) \text{ for some } n\in\mathbb{N}\\ 0 & : \text{otherwise}\end{cases}$$

The function f is uniformly locally Lipschitz because $\forall x\in X$, $B(x,\frac{\delta}{4})$ intersects at most one of the balls $B(x_m,\frac{\delta}{4})$ and f restricted to each ball $B(x_m,\frac{\delta}{4})$ is Lipschitz. We get a contradiction as f is unbounded. □

Remark 2.12. The equivalence of (a) and (f) of the previous theorem was proved in [Beer and Garrido (2014)] and the equivalence of the rest of the statements with (a) was proved in [Gupta and Kundu (2020)]. Note that from the previous theorem, it also follows that (X,d) is totally bounded if and only if each CC-regular function defined on it is bounded.

Our next result gives some more equivalent characterizations of totally bounded metric spaces using CC-regular and PC-regular functions.

Theorem 2.19. *[Gupta and Kundu (2020)] Let (Y,ρ) be a metric space. Then the following conditions are equivalent.*

- (a) *(Y,ρ) is totally bounded.*
- (b) *Whenever (X,d) is a metric space such that $(\widehat{X},d)$ has an accumulation point, then every function $f:(X,d)\to(Y,\rho)$ is PC-regular.*
- (c) *Whenever (X,d) is a metric space such that $(\widehat{X},d)$ has an accumulation point, then every function $f:(X,d)\to(Y,\rho)$ is CC-regular.*
- (d) *There exists a metric space (X,d) such that $(\widehat{X},d)$ has an accumulation point and every function $f:(X,d)\to(Y,\rho)$ is CC-regular.*

Proof. $(a)\Rightarrow(b)$: Let (X,d) be a metric space such that $(\widehat{X},d)$ has an accumulation point. Since Y is totally bounded, every sequence in (Y,ρ) is pseudo-Cauchy. Hence every function $f:(X,d)\to(Y,\rho)$ is PC-regular.

$(b)\Rightarrow(c)$: This follows from Proposition 1.7.

$(c)\Rightarrow(d)$: This is immediate.

$(d)\Rightarrow(a)$: Let (X,d) be a metric space such that $(\widehat{X},d)$ has an accumulation point and every function $f:(X,d)\to(Y,\rho)$ is CC-regular. Suppose (Y,ρ) is not totally bounded. Thus there exists a sequence (y_n) in Y and $\varepsilon>0$ such that $\rho(y_i,y_j)\geq\varepsilon$ for all $i,j\in\mathbb{N}$. Since the set of limit points of $(\widehat{X},d)$ is non-empty, there exists a Cauchy sequence of distinct terms $(\hat{x}_n)$ in $\hat{X}$. Thus there exists a sequence (x_n) in X consisting of distinct terms such that $d(x_n,\hat{x}_n)<1/n$ for all $n\in\mathbb{N}$. Thus, (x_n) is Cauchy in X. Define a function $f:(X,d)\to(Y,\rho)$ as follows:

$$f(x)=\begin{cases} y_n & : \; if\; x=x_n \text{ for some } n \\ y_1 & : \text{ otherwise}\end{cases}$$

According to the hypothesis, f should be CC-regular, but (x_n) is cofinally Cauchy in X such that $(f(x_n))$ is not cofinally Cauchy. We get a contradiction. Thus (Y,ρ) is totally bounded. □

Remark 2.13. Note that the condition, $(\widehat{X},d)$ has an accumulation point, is equivalent to saying that (X,d) has a Cauchy sequence with distinct terms.

Now we study how the isolation functional I and the local compactness functional ν are linked with the continuous CC-regular functions.

Theorem 2.20. *[Gupta and Kundu (2020)] Let f be a function from a metric space (X,d) to another metric space (Y,ρ). If for some $\varepsilon > 0$, f is uniformly continuous on $\{x \in X : I(x) < \varepsilon\}$, then f is continuous and CC-regular on X.*

Proof. Let (x_n) be a sequence of distinct points in X converging to a point $x \in X$. Then there exists $n_0 \in \mathbb{N}$ such that $d(x_n, x_m) < \frac{\varepsilon}{2}$ $\forall n, m \geq n_0$. Thus, f is uniformly continuous on the set $\{x\} \cup \{x_n : n \geq n_0\}$. But this implies that $(f(x_n))$ converges to $f(x)$. Hence the function is continuous.

Now suppose the function is not CC-regular. Therefore $\exists \varepsilon_o > 0$ such that $\forall n \in \mathbb{N}$, $\exists x_n \in X$ such that $f(B(x_n, \frac{1}{n}))$ cannot be contained in any finite union of open balls of radius ε_o in Y. Using the same technique as in Theorem 2.9, we can choose a sequence (z_n) in X consisting of infinite elements from each ball $B(x_n, \frac{1}{n})$ such that the image of the sequence is $\frac{\varepsilon_o}{2}$ discrete. Choose $\frac{1}{n_o} < \varepsilon$. Since the function is uniformly continuous on the set $A = \{x \in X : I(x) < \varepsilon\}$, there exists $\delta > 0$ such that $\forall x, y \in A$ with $d(x,y) < \delta$, $\rho(f(x), f(y)) < \frac{\varepsilon_o}{2}$. Choose $\frac{2}{m} < \min\{\frac{1}{n_o}, \delta\}$. Let $\{y,z\} \subset B(x_m, \frac{1}{m})$, where $y \neq z$ such that $\{y,z\} \subset \{z_n : n \in \mathbb{N}\}$. Now, $\{y,z\} \subset A$ with $d(y,z) < \delta$ but $\rho(f(y), f(z)) > \frac{\varepsilon_o}{2}$. We get a contradiction. □

Remark 2.14. The converse of the previous result may not be true. For example, the function f defined in Remark 2.8 is continuous and CC-regular, but there does not exist any $\varepsilon > 0$ such that f is uniformly continuous on $\{x \in X : I(x) < \varepsilon\}$.

Also, if we replace the isolation functional with the local compactness functional in the hypothesis of the previous result, then f need not be continuous. For example, take any discontinuous function from $\mathbb{R}$ to $\mathbb{R}$. For the local compactness functional, we have the following analogous result which can be proved in a manner similar to that of Theorem 2.20.

Theorem 2.21. *[Gupta and Kundu (2020)] Let f be a continuous function from a metric space (X,d) to another metric space (Y,ρ). If for some $\varepsilon > 0$, f is uniformly continuous on $\{x \in X : \nu(x) < \varepsilon\}$, then f is CC-regular on X.*

Theorem 2.22. *[Gupta and Kundu (2020)] Let (X,d) be a metric space. Suppose for every real-valued continuous function f defined on it, there exists some $\lambda > 0$ such that f is uniformly continuous on $\{x \in X : \nu(x) < \lambda\}$. Then (X,d) is cofinally complete.*

Proof. Let (x_n) be a cofinally Cauchy sequence of distinct points in X. By Proposition 1.1, there exists a pairwise disjoint family $\{\mathbb{M}_j : j \in \mathbb{N}\}$ of infinite subsets of $\mathbb{N}$ such that if $i \in \mathbb{M}_j$ and $l \in \mathbb{M}_j$ then $d(x_i, x_l) < \frac{1}{j}$. Suppose the sequence (x_n) does not cluster, thus the set $A = \{x_n : n \in \mathbb{N}\}$ is closed and discrete. Enumerate the elements of $\mathbb{M}_n$ as $l_1^n, l_2^n, l_3^n, \ldots$ for each $n \in \mathbb{N}$. Therefore, for each n, there exists unique i, $k \in \mathbb{N}$ such that $n = l_i^k$. Define a function f on A such that $\forall k \in \mathbb{N}$, $f(x_{l_i^k}) = 1$ for even values of i and $f(x_{l_i^k}) = 2$ for odd values of i. Since f is a continuous CC-regular function on A, extend it to a function F on X such that F is continuous and CC-regular.

We claim that there does not exist any $\lambda > 0$ such that F is uniformly continuous on $\{x \in X : \nu(x) < \lambda\}$. Suppose the claim is not true. Therefore $\exists \lambda > 0$ such that F is uniformly continuous on $\{x \in X : \nu(x) < \lambda\}$. Choose $\frac{1}{n_o} < \lambda$, consequently $\nu(x_k) < \frac{1}{n_o} < \lambda \; \forall x_k \in \bigcup\limits_{n \geqslant n_o} \mathbb{M}_n$. But F is not uniformly continuous on $\bigcup\limits_{n \geqslant n_o} \mathbb{M}_n$. We arrive at a contradiction. □

For additional reading, one may refer to [Burdick (2000); García-Máynez and Romaguera (1999)].

Exercises

Exercise 2.1

[Beer (2008)] Let (X,d) be a metric space. Then prove the following.

(a) $I(x) \leq \nu(x) \;\; \forall\, x \in X$
(b) $\nu(x) = \inf\{\liminf\limits_{n \to \infty} d(x, x_n) : (x_n) \text{ has no cluster point}\} \;\; \forall\, x \in X$.
(c) $A \subseteq B \Rightarrow \overline{\nu}(A) \leq \overline{\nu}(B)$
(d) $\overline{\nu}(A \cup B) = \max\{\overline{\nu}(A), \overline{\nu}(B)\}$
(e) $\overline{\nu}(A) = \overline{\nu}(cl(A))$
(f) $A \subseteq B \Rightarrow \underline{\nu}(A) \geq \underline{\nu}(B)$
(g) $\underline{\nu}(A \cup B) = \min\{\underline{\nu}(A), \underline{\nu}(B)\}$
(h) $\underline{\nu}(A) = \underline{\nu}(cl(A))$.

Exercise 2.2

Give an example of two equivalent metrics d and σ on X and a sequence (x_n) in X such that (x_n) is cofinally Cauchy in (X,d) but not in (X,σ).

Exercise 2.3

Illustrate that both the conditions given in Theorem 2.8(g) are important for the metric space (X,d) to be cofinally complete.

Exercise 2.4

Think of an example of a continuous function on a metric space (X,d) which is not uniformly locally bounded and hence verify that (X,d) is not cofinally complete by producing a cofinally Cauchy sequence in X which doesn't cluster in X.

Exercise 2.5

Prove Theorem 2.6, Proposition 2.1 and Theorem 2.10.

Exercise 2.6

[Beer *et al.* (2020)] Let (X,d) be a metric space. Then the following statements are equivalent.

(a) (X,d) is cofinally complete.
(b) Whenever $f:(X,d)\to\mathbb{R}$ is uniformly locally Lipschitz such that f is never zero, then $\frac{1}{f}$ is also uniformly locally Lipschitz.
(c) Whenever $f:(X,d)\to\mathbb{R}$ is Lipschitz such that f is never zero, then $\frac{1}{f}$ is uniformly locally Lipschitz.

Exercise 2.7

[Aggarwal and Kundu (2016)] Let (X,d) be a metric space. Then X is said to be *cofinally small* if either X is finite or $\forall\ \varepsilon>0$, $\exists\ x\in X$ and an infinite subset A of X such that $A\subseteq B(x,\varepsilon)$ (in other words, for every $\varepsilon>0$, there exists an infinite subset of X whose diameter is less than ε).

(a) Suppose $f:(X,d)\to(Y,\rho)$ is a one-to-one function. Then show that f is CC-regular if and only if $f(A)$ is a cofinally small subset of Y, for every cofinally small subset A of X.
(b) Prove that (X,d) is cofinally complete and discrete if and only if every cofinally small subset of (X,d) is finite.

Exercise 2.8

[Aggarwal and Kundu (2016)] Show that every cofinally complete, discrete subset of (X,d) is uniformly discrete if and only if every sequence (a_n) in X is either cofinally Cauchy or (a_n) satisfies the following condition: $\exists\ \varepsilon_o>0$ such that $\forall\ m,n\in\mathbb{N}\ (m\neq n)$, either $a_m=a_n$ or $d(a_m,a_n)>\varepsilon_o$.

Exercise 2.9

[Beer (2012)] Let (X,d) be a metric space. Let $F(x)=\inf\{\varepsilon>0:B(x,\varepsilon)$ is infinite$\}$ for $x\in X$. This functional $F(.)$ is called the *local finiteness functional.*

Moreover, (X,d) is said to be *strongly cofinally complete* if every sequence (x_n) in X with $\lim_{n\to\infty} F(x_n) = 0$ has a cluster point. Prove that the following are equivalent.

(a) (X,d) is strongly cofinally complete.
(b) The set of limit points X' of X is compact, and $\forall \varepsilon > 0$, $\exists \delta > 0$ such that whenever $d(x,X') > \varepsilon$, then $F(x) > \delta$.
(c) Whenever $\langle A_n \rangle$ is a decreasing sequence of non-empty closed subsets of X with $\lim_{n\to\infty} \sup\{F(x) : x \in A_n\} = 0$, then $\bigcap\{A_n : n \in \mathbb{N}\}$ is non-empty.
(d) Each sequence in X that is either cofinally Cauchy or lies in the set of imit points X' of X clusters.
(e) (X,d) is cofinally complete and X' is compact.
(f) Whenever (x_n) is a sequence in X with $\lim_{n\to\infty} \min\{\nu(x_n), d(x_n,X')\} = 0$, then (x_n) clusters.

Exercise 2.10

[Beer and Garrido (2014)] Let (X,d) be a metric space. Prove that the following statements are equivalent.

(a) (X,d) is totally bounded.
(b) Whenever (Y,ρ) is a metric space and $f : (X,d) \to (Y,\rho)$ is Cauchy-regular, then f is bounded.
(c) Whenever (Y,ρ) is a metric space and $f : (X,d) \to (Y,\rho)$ is uniformly locally Lipschitz, then f is bounded.

Exercise 2.11

[Beer and Garrido (2014)] Let (X,d) be a metric space. Prove that the following statements are equivalent.

(a) (X,d) is compact.
(b) Whenever (Y,ρ) is a metric space and $f : (X,d) \to (Y,\rho)$ is continuous, then f is bounded.
(c) Whenever (Y,ρ) is a metric space and $f : (X,d) \to (Y,\rho)$ is locally Lipschitz, then f is bounded.

Exercise 2.12

What about the converse of Theorem 2.22?

Chapter 3

Cofinal Completions

Since cofinally complete metric spaces are complete, it is natural to pay attention towards the metric spaces which have cofinally complete completion. In this chapter, we give various interesting characterizations of such spaces in terms of local total boundedness functional, Cauchy-regular functions, Cauchy-Lipschitz function, uniformly locally Lipschitz functions, etc. Furthermore, Cauchy-subregular functions are studied in detail.

3.1 Local Total Boundedness Functional

Recall the local compactness functional ν which was used to characterize cofinally complete metric spaces. Analogously, in this section we study a geometric functional, denoted by t, which measures the local total boundedness of a metric space at each point. We give examples to justify the use of the functional t over the local compactness functional in order to characterize the metric spaces having a cofinal completion.

Let us first recall an example of a metric space whose completion is not cofinally complete.

Example 3.1. Consider $X = \{e_n + \frac{1}{n}e_k : n,k \in \mathbb{N}\}$ as a metric subspace of the real Hilbert space ℓ_2, where $\{e_n : n \in \mathbb{N}\}$ is the standard orthonormal basis of ℓ_2. If we enumerate the elements of X, we will get a cofinally Cauchy sequence say (x_n) which has no cluster point. Thus (X,d) is not cofinally complete. Since (X,d) is complete, its completion is not cofinally complete.

Recall that Theorem 2.7 gives characterization of cofinally complete metric spaces using the local compactness functional. But this functional does not give an analogous characterization of metric spaces having a cofinal completion. Let us see an example.

Example 3.2. Let $X = \{n - \frac{1}{m} : m,n \in \mathbb{N}, n \geqslant 4, m \geqslant n\}$ equipped with the usual distance metric d. Evidently, the completion $(\widehat{X},d)$ of (X,d) is cofinally complete. Consider the sequence $(x_n) : 4-\frac{1}{5}, 5-\frac{1}{6}, 6-\frac{1}{7}, 7-\frac{1}{8}, \ldots$. It is clear that $\lim_{n\to\infty} \nu(x_n) = 0$ but the sequence (x_n) has no Cauchy subsequence and hence it does not cluster in $(\widehat{X},d)$.

One may also consider the functional $\widehat{\nu}$, which denotes the local compactness functional on $(\widehat{X},d)$. But in order to study some internal conditions on any metric space itself, we consider the following geometric functional that measures the local total boundedness of a metric space at each point. This functional plays a key role in the study of metric spaces having cofinal completion.

Definition 3.1. Let (X,d) be a metric space and $x \in X$. The functional $t(\cdot)$ is defined as follows: if x has no totally bounded neighborhood, set $t(x) = 0$, otherwise, put $t(x) = sup\{\varepsilon > 0 : B_d(x,\varepsilon) \text{ is totally bounded}\}$. This geometric functional is called the *local total boundedness functional* on X. The set $\{x \in X : t(x) = 0\}$ is the set of points of non-local total boundedness of X, which is denoted by $nlt(X)$. Thus a metric space (X,d) is said to be *locally totally bounded* if $t(x) > 0\ \forall x \in X$, while it is called *uniformly locally totally bounded* if $\inf\{t(x) : x \in X\} > 0$.

Let us first collect some elementary properties of this defined functional.

Theorem 3.1. *Let (X,d) be a metric space. Then for each $x \in X$, $\widehat{\nu}(x) = t(x)$, that is, $\widehat{\nu}|_X = t$.*

Proof. Let $x \in X$. First we will prove that $\widehat{\nu}(x) \leq t(x)$. If $t(x) = 0$, then clearly, $\widehat{\nu}(x) > 0$ is not possible. If $t(x) = \alpha > 0$, then suppose $\widehat{\nu}(x) = \beta > \alpha$. Since $t(x) = \alpha$, $B(x,\gamma)$, where $\gamma = \alpha + \frac{\beta-\alpha}{2}$, contains a sequence which does not have any Cauchy subsequence. Thus $\widehat{\nu}(x) \leq \gamma < \beta$, a contradiction. Thus $\widehat{\nu}(x) \leq t(x)$. To see that $t(x) \leq \widehat{\nu}(x)$, first note that $\widehat{\nu}(x) = 0$ implies $t(x) = 0$. Next, for the case $\widehat{\nu}(x) = \alpha > 0$, suppose $t(x) = \beta > \alpha$. Then the completion of $B(x, \alpha + \frac{\beta-\alpha}{2})$ would be compact, which implies $\widehat{\nu}(x) > \alpha$, a contradiction. Hence $\widehat{\nu}(x) = t(x)\ \forall x \in X$. □

Corollary 3.1. *Let (X,d) be a complete metric space. Then $\nu(x) = t(x)$ for all $x \in X$.*

Considering the metric space defined in Example 3.2, one can see that in general, $\nu(x) \neq t(x)$ for each $x \in X$. In the previous corollary we have observed that $\nu(x) = t(x)\ \forall x \in X$ if (X,d) is complete. The next example shows that the converse need not be true.

Example 3.3. Let $X = \{e_n, e_n + \frac{1}{n}e_k : n,k \in \mathbb{N}\}$. We define a metric d on X as follows:

- $d(e_n, e_m) = \left|\frac{2}{n} - \frac{2}{m}\right| \ \forall n \neq m \in \mathbb{N}$.
- $d(e_n, e_n + \frac{1}{n}e_k) = \frac{1}{n} \ \forall n,k \in \mathbb{N}$.
- $d(e_n, e_m + \frac{1}{m}e_k) = \left|\frac{2}{n} - \frac{2}{m}\right| + \frac{1}{m} \ \forall n \neq m$ and $\forall k \in \mathbb{N}$.
- $d(e_n + \frac{1}{n}e_k, e_n + \frac{1}{n}e_m) = \frac{\sqrt{2}}{n} \ \forall k \neq m$ and $\forall n \in \mathbb{N}$.
- $d(e_n + \frac{1}{n}e_k, e_m + \frac{1}{m}e_t) = \left|\frac{2}{n} - \frac{2}{m}\right| + \frac{1}{n} + \frac{1}{m} \ \forall n \neq m$ and $\forall k,t \in \mathbb{N}$.

The sequence (e_n) is a Cauchy sequence but without any cluster point. Thus (X,d) is not complete. Also, it can be verified that $\nu(x) = t(x)$ for each $x \in X$.

Lemma 3.1. *Let (X,d) be a metric space. Suppose there exists a bounded subset of X which is not totally bounded, then $t : (X,d) \to [0,\infty)$ is uniformly continuous.*

Proof. The result follows from Lemma 2.1 and Theorem 3.1. □

Let us start with the following characterization using the functional t, which will further be used to give more characterizations of metric spaces having cofinal completion using Cauchy-regular functions.

Theorem 3.2. *[Gupta and Kundu (2021b)] Let (X,d) be a metric space. Then the following statements are equivalent.*

(*a*) *The completion $(\widehat{X},d)$ of (X,d) is cofinally complete.*
(*b*) *If (x_n) is a sequence in X with $\lim_{n\to\infty} t(x_n) = 0$, then (x_n) has a Cauchy subsequence.*

Proof. $(a) \Rightarrow (b)$: Let (x_n) be a sequence in X with $\lim_{n\to\infty} t(x_n) = 0$. Thus by Theorem 3.1, $\lim_{n\to\infty} \hat{\nu}(x_n) = 0$ in $(\widehat{X},d)$. Consequently, by Theorem 2.7, (x_n) has a cluster point in $(\widehat{X},d)$ and hence the implication follows.

$(b) \Rightarrow (a)$: Suppose $(\hat{x}_n)$ is a cofinally Cauchy sequence of distinct elements in $(\widehat{X},d)$ with no Cauchy subsequence. Then there exists a sequence (x_n) of distinct elements in X such that $d(\hat{x}_n, x_n) < \frac{1}{n} \ \forall n \in \mathbb{N}$. Consequently, (x_n) is cofinally Cauchy in (X,d) with no Cauchy subsequence. For $\varepsilon = 1$, there exists an infinite subset $\mathbb{N}_1$ of $\mathbb{N}$ such that $d(x_n,x_m) < 1$ for all $n,m \in \mathbb{N}_1$. Choose $n_1 \in \mathbb{N}_1$. Thus $t(x_{n_1}) < 1$. For $\varepsilon = \frac{1}{2}$, there exists an infinite subset $\mathbb{N}_2$ of $\mathbb{N}$ such that $d(x_n,x_m) < \frac{1}{2}$ for all $n,m \in \mathbb{N}_2$. Choose $n_2 \in \mathbb{N}_2$ such that $n_2 > n_1$, thus $t(x_{n_2}) < \frac{1}{2}$. Thus we will get a sequence (x_{n_k}) such that $\lim_{k\to\infty} t(x_{n_k}) = 0$, but (x_{n_k}) has no Cauchy subsequence, a contradiction. □

3.2 Cofinal Completion vis-à-vis Cauchy-regular Functions

In this section, we characterize the metric spaces having cofinal completion in terms of Cauchy-regular functions and some functions that are stronger than the Cauchy-regular functions, namely Cauchy-Lipschitz functions and uniformly locally Lipschitz functions.

Our first set of characterizations includes the condition under which every Cauchy-regular function is CC-regular. Observe that a Cauchy-regular function need not be CC-regular.

Example 3.4. Consider the metric space (X,d) defined in Example 3.1. The function $f : X \to \mathbb{N}$ defined by, $f(x_n) = n$, is Cauchy-regular, but it is not CC-regular.

Let us also recall from [Williamson and Janos (1987)] that two metrics on a set are said to be *Cauchy equivalent* if the collections of Cauchy sequences with respect to both the metrics are same. Clearly, Cauchy equivalence is an equivalence relation on the collection of all metrics on a set X. Moreover, metrics d and ρ on a set X are Cauchy equivalent if and only if the identity functions between the two metric spaces, (X,d) and (X,ρ), are Cauchy-regular. Evidently, all Cauchy equivalent metrics on a set are equivalent. But the converse is not in general true. For example, consider $\mathbb{N}$ with the metrics d and ρ defined as: $d(n,m) = |n-m|$ and $\rho(n,m) = |\frac{1}{n} - \frac{1}{m}|$.

The next proposition gives a way to construct different Cauchy equivalent metrics on a set.

Proposition 3.1. *Let (X,d) be a metric space and let f be a function on (X,d) with values in another metric space (Y,ρ). Define a metric σ on X by:*

$$\sigma(a,b) = d(a,b) + \rho(f(a), f(b)) \text{ for } a, b \text{ in } X$$

Then the metrics d and σ are Cauchy equivalent on X if and only if f is Cauchy-regular.

Here is an interesting characterization of Cauchy equivalent metrics.

Proposition 3.2. *Two metrics d and ρ on a set X are Cauchy equivalent if and only if there is a homeomorphism between their completions that fixes X.*

Proof. Suppose there exists a homeomorphism between the completions $(\widehat{X}_d, d)$ and $(\widehat{X}_\rho, \rho)$ of (X,d) and (X,ρ) respectively, that fixes X. Since every continuous function on a complete space is Cauchy-regular, the metrics d and ρ are Cauchy equivalent on X.

Conversely, suppose that the metrics d and ρ are Cauchy equivalent on X. Then the identity map $id : (X,d) \to (X,\rho)$ is Cauchy-regular. Let $x \in \widehat{X}_d$. Then there exists a sequence (x_n) in X which converges to x with respect to the metric d and hence the sequence (x_n) converges to some x' in $(\widehat{X}_\rho,\rho)$. Suppose (z_n) is another sequence in X that converges to x with respect to the metric d. Then there exists some z in $\widehat{X}_\rho$ such that (z_n) converges to z with respect to the metric ρ. Now, the sequence $(x_1,\, z_1,\, x_2,\, z_2,\ldots)$ is convergent to x in $(\widehat{X}_d,d)$ and hence the sequence is convergent in $(\widehat{X}_\rho,\rho)$. So $x' = z$. Consequently, the function $f : (\widehat{X}_d,d) \to (\widehat{X}_\rho,\rho) :\ f(x) = x'$ is well-defined, that is, $f(x)$, for $x \in \widehat{X}_d$, is independent of the choice of the sequence in X converging to x with respect to d. Note that f is a continuous extension of id. Moreover, f is a homeomorphism. □

Remark 3.1. Note that if d and ρ are Cauchy equivalent metrics on X, then the corresponding completions need not be isometric. For example, let $X = (0,2)$ and d be the usual distance metric. Define a metric ρ as: $\rho(a,b) = |a-b| + |a^2 - b^2|$ for $a,\ b \in X$. Then by Proposition 3.1, d and ρ are Cauchy equivalent on X. Moreover, $([0,2],d)$ and $([0,2],\rho)$ are the completions of $((0,2),d)$ and $((0,2),\rho)$ respectively. But they are not isometric, because $\rho(0,2) = 6$ and $|x-y| \le 2$ for all $x,\ y \in [0,2]$.

Now we are ready to study the following set of characterizations of metric spaces having cofinal completion.

Theorem 3.3. *Let (X,d) be a metric space. Then the following statements are equivalent.*

(a) *The completion $(\widehat{X},d)$ of (X,d) is cofinally complete.*
(b) *For every metric σ on X Cauchy equivalent to d and each $\varepsilon > 0$, there exists $\delta > 0$ such that $\forall\, x \in X,\ B_d(x,\delta) \subseteq \bigcup_{i=1}^{n} B_\sigma(x_i,\varepsilon)$ for some finite subset $\{x_1,\ldots,x_n\}$ of X.*
(c) *Every d-cofinally Cauchy sequence in X is σ-cofinally Cauchy for all Cauchy equivalent metrics σ on X.*
(d) *Every Cauchy-regular function on (X,d) with values in a metric space (Y,ρ) is CC-regular.*
(e) *Every Cauchy-Lipschitz function on (X,d) with values in a metric space (Y,ρ) is CC-regular.*
(f) *Every cofinally Cauchy sequence in (X,d) has a Cauchy subsequence.*
(g) *Every complete subset (as a metric subspace) of (X,d) is cofinally complete.*
(h) *Every complete and locally compact subset of (X,d) is uniformly locally compact in its relative topology.*

(i) *For every Cauchy equivalent metric σ, there exists $\delta > 0$ such that for all $x \in X$, $B_d(x,\delta)$ is a σ-bounded set.*
(j) *Every Cauchy-regular function on (X,d) with values in a metric space (Y,ρ) is uniformly locally bounded.*
(k) *Every Cauchy-Lipschitz function on (X,d) with values in a metric space (Y,ρ) is uniformly locally bounded.*

Proof. The implications $(d) \Rightarrow (e)$, $(f) \Rightarrow (g)$ and $(j) \Rightarrow (k)$ are immediate, whereas the implication $(g) \Rightarrow (h)$ follows from Theorem 2.8.

$(a) \Rightarrow (b)$: Let σ be a metric on X Cauchy equivalent to d. Then the identity map $id : (X,d) \to (X,\sigma)$ is Cauchy-regular. So there exists a continuous function $f' : (\widehat{X}_d, d) \to (\widehat{X}_\sigma, \sigma)$ which extends id, where $(\widehat{X}_d, d)$ and $(\widehat{X}_\sigma, \sigma)$ denote the completions of (X,d) and (X,σ) respectively. Now, the metric σ' defined on $\widehat{X}_d$ as:

$$\sigma'(a,b) = d(a,b) + \sigma(f'(a), f'(b)) \text{ for } a, b \in \widehat{X}_d$$

is equivalent to d on $\widehat{X}_d$. Let $\varepsilon > 0$. Then by Theorem 2.8, we get the required δ.

$(b) \Rightarrow (c)$: Similar to the proof of the implication $(b) \Rightarrow (c)$ of Theorem 2.8.

$(c) \Rightarrow (d)$: Let $f : (X,d) \to (Y,\rho)$ be a Cauchy-regular function. Now, define a metric σ on X as follows:

$$\sigma(a,b) = d(a,b) + \rho(f(a), f(b)) \text{ for } a,\ b \text{ in } X$$

Then σ is Cauchy equivalent to d by Proposition 3.1. Let (x_n) be a d-cofinally Cauchy sequence in X, then it is σ-cofinally Cauchy. Hence by the definition of σ, $(f(x_n))$ is cofinally Cauchy in (Y,ρ).

$(e) \Rightarrow (f)$: Since a cofinally Cauchy sequence with no constant subsequence has a cofinally Cauchy subsequence of distinct terms, it is enough to prove that every cofinally Cauchy sequence of distinct points in (X,d) has a Cauchy subsequence. Let (x_n) be such a sequence in X. Suppose that (x_n) has no Cauchy subsequence. Consider the function $f : A \to \mathbb{R}$: $f(x_n) = n$, where $A = \{x_n : n \in \mathbb{N}\}$. We claim that f is Cauchy-regular. Let (y_m) be a Cauchy sequence of distinct points in A. Then $y_m = x_n$ for some n. Now, for $\varepsilon = 1$, $\exists\ p_1 \in \mathbb{N}$ such that $d(y_j, y_k) < 1$ $\forall\ j,\ k \geq p_1$. Let $A_1 = \{n \in \mathbb{N} : x_n = y_j \text{ for some } j \geq p_1\}$, then $d(x_n, x_m) < 1$ $\forall\ n,\ m \in A_1$. Similarly, for $\varepsilon = \frac{1}{2}$, $\exists\ p_2 \in \mathbb{N}$ $(p_2 > p_1)$ such that $d(y_j, y_k) < \frac{1}{2}$ $\forall\ j,\ k \geq p_2$. Let $A_2 = \{n \in A_1 : x_n = y_j \text{ for some } j \geq p_2\}$ then $d(x_n, x_m) < \frac{1}{2}$ $\forall\ n,\ m \in A_2$. By induction, we get $A_k = \{n \in A_{k-1} : x_n = y_j \text{ for some } j \geq p_k\}$ where $p_k > p_{k-1}$ and $d(x_n, x_m) < \frac{1}{k}$ $\forall\ n,\ m \in A_k$. Choose any $n_1 \in A_1$. Now, for $k \in \mathbb{N} \setminus \{1\}$, choose $n_k \in A_k$ such that $n_k > n_{k-1}$. Then $(x_{n_k})_{k \in \mathbb{N}}$ is a Cauchy subsequence of (x_n). We get a contradiction. Thus, there does not exist any Cauchy sequence of distinct points in A and hence any Cauchy sequence in A will be eventually constant. Consequently, f is Cauchy-regular. By Corollary 1.1, f can be extended to a real-valued Cauchy-regular function f' on (X,d). Then by Theorem 1.5, for $0<\varepsilon<1$, there exists a Cauchy-Lipschitz function $g:(X,d)\to\mathbb{R}$ with $\sup_{x\in X}|f'(x)-g(x)|\leq\frac{\varepsilon}{3}$. By (e), g is CC-regular. Since (x_n) is cofinally Cauchy, $(g(x_n))$ is also cofinally Cauchy. Consequently, there exists an infinite subset N of $\mathbb{N}$ such that $|g(x_n)-g(x_m)|<\frac{\varepsilon}{3}\ \forall\, n,\, m\in N$. Hence

$$\begin{aligned}1\leq|n-m|&=|f(x_n)-f(x_m)|\\&\leq|f(x_n)-g(x_n)|+|g(x_n)-g(x_m)|+|g(x_m)-f(x_m)|\\&<\varepsilon<1\quad\forall\, n,\, m\in N,\ n\neq m.\end{aligned}$$

We arrive at a contradiction. Thus every cofinally Cauchy sequence in (X,d) has a Cauchy subsequence.

$(h)\Rightarrow(a)$: Let $(\widehat{x}_n)$ be a cofinally Cauchy sequence in $(\widehat{X},d)$ without cluster point. This sequence cannot have a Cauchy subsequence. Then there exists a sequence (x_n) in X such that $d(x_n,\widehat{x}_n)<\frac{1}{n}$ for all $n\in\mathbb{N}$. Consequently, (x_n) is cofinally Cauchy in (X,d) with no Cauchy subsequence and hence the set $A=\{x_n:n\in\mathbb{N}\}$ is complete and discrete. By (h), A is uniformly locally compact and since (x_n) is cofinally Cauchy, it has a convergent subsequence. Hence we get a contradiction. Thus, the sequence $(\widehat{x}_n)$ has a Cauchy subsequence and hence it has a cluster point in $\widehat{X}$.

$(a)\Rightarrow(i)$: Let σ be a metric on X Cauchy equivalent to d. Then the identity map $id:(X,d)\to(X,\sigma)$ is Cauchy-regular and thus there exists a continuous function $f':(\widehat{X}_d,d)\to(\widehat{X}_\sigma,\sigma)$ which extends id. By Theorem 2.8, f' is uniformly locally bounded and hence id is uniformly locally bounded. Subsequently, there exists $\delta>0$ such that for all $x\in X$, $B_d(x,\delta)$ is a σ-bounded set.

$(i)\Rightarrow(j)$: Similar to the proof of the implication $(c)\Rightarrow(d)$.

$(k)\Rightarrow(a)$: We prove it by using Theorem 2.8. Let $f:(\widehat{X},d)\to\mathbb{R}$ be a locally Lipschitz function. We claim that f is uniformly locally bounded. Since $(\widehat{X},d)$ is complete, f is Cauchy-Lipschitz by Theorem 1.3. This implies $f|_X$ is Cauchy-Lipschitz and hence uniformly locally bounded. Suppose there exists $\delta>0$ such that $\forall\, x\in X,\ f(B(x,\delta)\cap X)$ is bounded. Let $\widehat{x}\in\widehat{X}$. We claim that $f(B(\widehat{x},\frac{\delta}{3}))$ is bounded. Suppose it is not bounded. Then there exists a sequence $(\widehat{x}_n)$ in $B(\widehat{x},\frac{\delta}{3})$ such that $|f(\widehat{x}_n)|>n$. Let $(a_m^n)_{m\in\mathbb{N}}$ be a sequence in X converging to $\widehat{x}_n$ for every $n\in\mathbb{N}$. Thus, for all $n\in\mathbb{N}$ there exists $m_n\in\mathbb{N}$ such that $d(a_k^n,\widehat{x}_n)<\frac{\delta}{3}\ \forall\, k\geq m_n$. Now let $x\in X$ such that $d(x,\widehat{x})<\frac{\delta}{3}$. Then $d(a_k^n,x)<\delta\ \ \forall\, k\geq m_n$ and $\forall\, n\in\mathbb{N}$. Consequently, there exists $M>0$ such that $|f(a_k^n)|<M\ \ \forall\, k\geq m_n$ and $\forall\, n\in\mathbb{N}$. By the continuity of f, $(f(\widehat{x}_n))$ is bounded. We get a contradiction. Thus f is uniformly locally bounded. □

Remark 3.2. (i) The conditions (d), (f), (g) and (j) were proved to be equivalent to (a) in [Aggarwal and Kundu (2016)].
(ii) Theorem 3.3 still holds if we replace the metric space (Y,ρ) by $(\mathbb{R},|\cdot|)$.

Recall from Chapter 1 that the class of real-valued locally Lipschitz functions on an arbitrary metric space is uniformly dense in the class of continuous functions, while the class of real-valued Lipschitz in the small functions is uniformly dense in the class of uniformly continuous functions. Before the introduction of Cauchy-Lipschitz functions, it was naturally guessed that the collection of real-valued uniformly locally Lipschitz functions should be uniformly dense in the collection of Cauchy-regular functions. But in [Beer and Garrido (2015)], it was proved that this is not in general true; in fact it is true if the completion of the domain of the function is cofinally complete. Later on, in [Beer and Garrido (2016)], a new class of functions namely Cauchy-Lipschitz was defined which played the role parallel to Cauchy-regular functions in Lipschitz environment. Moreover, they proved that the cofinal completeness of the completion of a metric space (X,d) is equivalent to the condition that every Cauchy-Lipschitz function from (X,d) to an arbitrary metric space is uniformly locally Lipschitz. In order to make our work self-contained, we prove the aforesaid results of [Beer and Garrido (2015)] and [Beer and Garrido (2016)].

Theorem 3.4. *Let (X,d) be a metric space. Then the following statements are equivalent :*

(a) *The completion $(\widehat{X},d)$ of (X,d) is cofinally complete.*
(b) *Every Cauchy-Lipschitz function on (X,d) with values in an arbitrary metric space (Y,ρ) is uniformly locally Lipschitz.*
(c) *Whenever $(Y,||\cdot||)$ is a Banach space and $f:X\to Y$ is Cauchy-regular, then f can be uniformly approximated by uniformly locally Lipschitz functions.*

Proof. $(a)\Rightarrow(b)$: Let $f:(X,d)\to(Y,\rho)$ be a Cauchy-Lipschitz function and let $x\in\widehat{X}$. Then there exists a sequence (x_n) in X which converges to x and hence the sequence $(f(x_n))$ converges to some y in $(\widehat{Y},\rho)$. Now the function $\widehat{f}:(\widehat{X},d)\to(\widehat{Y},\rho)$: $\widehat{f}(x)=y$ is well-defined, that is, $\widehat{f}(x)$, for $x\in\widehat{X}$, is independent of the choice of the sequence in X converging to x. Note that $\widehat{f}$ is a locally Lipschitz function such that $\widehat{f}(z)=f(z)$ for $z\in X$. By Theorem 2.8, the function $\widehat{f}$ and hence f is uniformly locally Lipschitz.

$(b)\Rightarrow(c)$: It follows from Theorem 1.5.

$(c) \Rightarrow (a)$: We prove it by using Theorem 3.3. Let $f : (X,d) \to \mathbb{R}$ be a Cauchy-regular function. Then by (c), given $\varepsilon > 0$, there exists a uniformly locally Lipschitz function $g : (X,d) \to \mathbb{R}$ with $\sup_{x\in X} |f(x) - g(x)| \leq \varepsilon$. Hence the function f is uniformly locally bounded as g is so. Consequently by Theorem 3.3, $(\widehat{X},d)$ is cofinally complete. □

Remark 3.3. The previous theorem still holds if we replace Y by $(\mathbb{R}, |\cdot|)$.

Recall that in Theorem 2.8, we have studied the ν-boundedness of real-valued continuous functions. Here, we study the t-boundedness of Cauchy-regular functions, Cauchy-Lipschitz functions and uniformly locally Lipschitz functions and characterize metric spaces having cofinal completion.

Definition 3.2. Let (X,d) be a metric space. A real-valued function f defined on (X,d) is said to be *t-bounded* if there exists $r > 0$ such that $\{f(x) : x \in X,\ t(x) < r\}$ is bounded, while f is said to be t-bounded on a non-empty subset A of X if there exists $r > 0$ such that the set $\{f(x) : x \in A,\ t(x) < r\}$ is bounded.

It can be easily verified that a t-bounded function need not be a bounded function.

Theorem 3.5. *[Gupta and Kundu (2021b)] Let (X,d) be a metric space. Then the following statements are equivalent.*

- (a) *The completion $(\widehat{X},d)$ of (X,d) is cofinally complete.*
- (b) *Let f be any real-valued Cauchy-regular function on (X,d). Then there exists $n_0 \in \mathbb{N}$ such that every point of the set $A = \{x : |f(x)| \geq n_0\}$ has a totally bounded neighbourhood. Moreover, $\inf\{t(x) : x \in A\} > 0$.*
- (c) *Every real-valued Cauchy-regular function on (X,d) is t-bounded.*
- (d) *Every real-valued Cauchy-Lipschitz function on (X,d) is t-bounded.*
- (e) *Every real-valued uniformly locally Lipschitz function on (X,d) is t-bounded.*
- (f) *Every real-valued uniformly locally Lipschitz function on (X,d) is t-bounded on every non-empty proper subset of X.*

Proof. The implications $(c) \Rightarrow (d) \Rightarrow (e) \Rightarrow (f)$ are all immediate.

$(a) \Rightarrow (b)$: Let f be any real-valued Cauchy-regular function on (X,d). By Corollary 1.1, we can extend it to a function $\hat{f} : \widehat{X} \to \mathbb{R}$ such that $\hat{f}$ is Cauchy-regular and hence continuous. Since $(\widehat{X},d)$ is cofinally complete, by Theorem 2.8, there exists n_o such that for $A' = \{x \in \widehat{X} : |\hat{f}(x)| \geq n_o\}$, $\inf\{\hat{\nu}(x) : x \in A'\} > 0$. Now let $A = \{x : |f(x)| \geq n_o\}$. It is clear that $A \subseteq A'$. Thus by Theorem 3.1, we are done.

$(b) \Rightarrow (c)$: Let $f : (X,d) \to \mathbb{R}$ be a Cauchy-regular function. Thus, $\exists n_o \in \mathbb{N}$ such that for $A = \{x \in X : |f(x)| \geq n_o\}$, $\inf\{t(x) : x \in A\} > 0$. Let $\inf\{t(x) : x \in A\} = r$. Note that the set $\{f(x) : x \in X,\ t(x) < r\}$ is bounded because $t(x) < r$, implies that $x \notin A$, which further shows that $|f(x)| < n_o$. Hence f is t-bounded.

$(f) \Rightarrow (a)$: Suppose there exists a cofinally Cauchy sequence $(\hat{x}_n)$ in $(\widehat{X},d)$ with no cluster point. Then there exists a sequence (x_n) in X such that $d(\hat{x}_n, x_n) < \frac{1}{n}\ \forall n \in \mathbb{N}$. Consequently, (x_n) is cofinally Cauchy in (X,d) with no Cauchy subsequence. By passing to a subsequence, we see that there exists a $\delta > 0$ for which the family of open balls $\{B(x_n, \delta) : n \in \mathbb{N}\}$ is pairwise disjoint. Define a function $f : (X,d) \to \mathbb{R}$ as follows:

$$f(x) = \begin{cases} n\left(1 - \frac{4}{\delta} d(x,x_n)\right) & : x \in B\left(x_n, \frac{\delta}{4}\right) \text{ for some } n \in \mathbb{N} \\ 0 & : \text{otherwise} \end{cases}$$

The function f is uniformly locally Lipschitz because $\forall x \in X$, $B(x, \frac{\delta}{4})$ intersects at most one of the balls $B(x_m, \frac{\delta}{4})$ and f restricted to each ball $B(x_m, \frac{\delta}{4})$ is Lipschitz. Let $B = A \setminus \{x_1\}$. Thus, $\exists r > 0$ such that the set $\{f(x) : x \in B,\ t(x) < r\}$ is bounded. Therefore, $\exists M > 0$ such that $|f(x)| \leq M\ \forall x \in B$ such that $t(x) < r$. Since (x_n) is a cofinally Cauchy sequence, there exists an infinite subset $\mathbb{N}_r$ of $\mathbb{N}$ such that $d(x_n, x_m) < r\ \ \forall n, m \in \mathbb{N}_r$, which implies $t(x_n) < r\ \forall n \in \mathbb{N}_r$. Hence, $|f(x_n)| \leq M\ \forall n \in \mathbb{N}_r$. Since (x_n) is a sequence of distinct points and $\mathbb{N}_r$ is an infinite set, we arrive at a contradiction. □

Since it is known that every Cauchy-regular function on a metric space (X,d) is uniformly continuous if and only if $(\widehat{X},d)$ is UC [Beer (1986); Jain and Kundu (2007)], our next aim is to give a collection of functions which are stronger than Cauchy-regular functions so that an analogous result holds for cofinal completion. In this regard, we define the following set which is analogous to the one mentioned in Theorem 2.10.

Definition 3.3. Let (X,d) and (Y,ρ) be metric spaces. Then $f \in FV(X,Y)$ if f is Cauchy-regular and for all $\varepsilon > 0, f$ is uniformly continuous on the set $\{x \in X : t(x) > \varepsilon\}$.

Before stating the main result, let us first study the following lemma which gives some relation between the sets $CV(X,Y)$ and $FV(X,Y)$. In the proof of the lemma, we make use of a well-known fact about asymptotic sequences: a function $f : (X,d) \to (Y,\rho)$ is uniformly continuous if and only if, for sequences (x_n) and (y_n) in X with $(x_n) \asymp (y_n)$, we have $(f(x_n)) \asymp (f(y_n))$. Recall that two sequences (x_n) and (y_n) in a metric space (X,d) are said to be asymptotic, that is, $(x_n) \asymp (y_n)$, if for every $\varepsilon > 0$, there exists $n_o \in \mathbb{N}$ such that $d(x_n, y_n) < \varepsilon$ for all $n, m \geq n_o$.

Lemma 3.2. *[Gupta and Kundu (2021b)] Let (X,d) and (Y,ρ) be metric spaces. Then,*

(a) *If $g \in CV(\widehat{X},Y)$, then $g|_X \in FV(X,Y)$.*
(b) *Let $g \in FV(X,Y)$. If $\widehat{g} : (\widehat{X},d) \to (\widehat{Y},\rho)$ is the continuous extension of g, then $\widehat{g} \in CV(\widehat{X},\widehat{Y})$.*

Proof. (a): Let $g \in CV(\widehat{X},Y)$. Since $(\widehat{X},d)$ is complete, g is Cauchy-regular. Let $\varepsilon > 0$, $\{x \in X : \iota(x) > \varepsilon\} = \{x \in \widehat{X} : \hat{\nu}(x) > \varepsilon\} \cap X$. Since g is uniformly continuous on the set $\{x \in \widehat{X} : \hat{\nu}(x) > \varepsilon\}$, g is uniformly continuous on the set $\{x \in X : \iota(x) > \varepsilon\}$. Thus $g|_X \in FV(X,Y)$.

(b): Let $\lambda > 0$. We need to prove that $\widehat{g}$ is uniformly continuous on the set $A = \{x \in \widehat{X} : \hat{\nu}(x) > \lambda\}$. Let $(x_n) \asymp (y_n)$ such that $\{x_n, y_n : n \in \mathbb{N}\} \subseteq A$. We claim that $(\widehat{g}(x_n)) \asymp (\widehat{g}(y_n))$. Let $\varepsilon > 0$ and $n \in \mathbb{N}$ be fixed. Since $\widehat{g}$ is continuous at x_n, $\exists\, \delta_{x_n} > 0$ such that $d(x_n,z) < \delta_{x_n}$ implies $\rho(\widehat{g}(x_n),\widehat{g}(z)) < \frac{\varepsilon}{3}$. Let δ_{y_n} be the corresponding value for (y_n). Now the uniform continuity of the functional ν implies that $\exists\, \delta > 0$ such that if $d(x,y) < \delta$, then $|\nu(x) - \nu(y)| < \frac{\lambda}{2}$. Since X is dense in $\widehat{X}$, we can choose a sequence (x'_n) in X such that for each n, $d(x_n,x'_n) < \min\{\frac{1}{n}, \delta, \delta_{x_n}\}$. Similarly, choose a sequence (y'_n) in X such that for each n, $d(y_n,y'_n) < \min\{\frac{1}{n}, \delta, \delta_{y_n}\}$. By construction, it is clear that $\{x'_n, y'_n : n \in \mathbb{N}\} \subseteq B = \{x \in X : \iota(x) > \frac{\lambda}{2}\}$ and by usual arguments, one can see that $(x'_n) \asymp (y'_n)$. Since g is uniformly continuous on B, we have $(g(x'_n)) \asymp (g(y'_n))$, that is, $(\widehat{g}(x'_n)) \asymp (\widehat{g}(y'_n))$. Thus there exists $n_o \in \mathbb{N}$ such that $\forall n \geqslant n_o$, $\rho(\widehat{g}(x'_n),\widehat{g}(y'_n)) < \frac{\varepsilon}{3}$. Now, for all $n \geqslant n_o$,

$$\begin{aligned}
\rho(\widehat{g}(x_n),\widehat{g}(y_n)) &< \rho(\widehat{g}(x_n),\widehat{g}(x'_n)) + \rho(\widehat{g}(x'_n),\widehat{g}(y'_n)) + \rho(\widehat{g}(y'_n),\widehat{g}(y_n)) \\
&< \frac{\varepsilon}{3} + \frac{\varepsilon}{3} + \frac{\varepsilon}{3} \\
&= \varepsilon
\end{aligned}$$

Hence $\widehat{g}$ is uniformly continuous on A and consequently, $\widehat{g} \in CV(\widehat{X},\widehat{Y})$. □

Theorem 3.6. *[Gupta and Kundu (2021b)] Let (X,d) be a metric space. Then the following statements are equivalent.*

(a) *The completion $(\widehat{X},d)$ of (X,d) is cofinally complete.*
(b) *Whenever (Y,ρ) is a metric space and $f \in FV(X,Y)$, then f is uniformly continuous on (X,d).*
(c) *Whenever (Y,ρ) is a metric space and f is a Cauchy-Lipschitz function belonging to $FV(X,Y)$, then f is uniformly continuous on (X,d).*
(d) *Whenever (Y,ρ) is a metric space and f is a uniformly locally Lipschitz function belonging to $FV(X,Y)$, then f is uniformly continuous on (X,d).*

(*e*) *If f is a bounded uniformly locally Lipschitz function belonging to $FV(X,\mathbb{R})$, then f is uniformly continuous on (X,d).*

Proof. The implications $(b) \Rightarrow (c) \Rightarrow (d) \Rightarrow (e)$ are all immediate.

$(a) \Rightarrow (b)$: Let (Y,ρ) be a metric space and $f \in FV(X,Y)$. By Lemma 3.2, $\widehat{f} \in CV(\widehat{X},\widehat{Y})$. Since $(\widehat{X},d)$ is cofinally complete, by Theorem 2.10, $\widehat{f}$ is uniformly continuous on $(\widehat{X},d)$ and thus f is uniformly continuous on (X,d).

$(e) \Rightarrow (a)$: Suppose $(\widehat{X},d)$ is not cofinally complete. Therefore, by Theorem 3.2, there exists a sequence (x_n) in X such that $\lim_{n\to\infty} t(x_n) = 0$, but it has no Cauchy subsequence. If needed, by passing to a subsequence, we see that there exists a $\delta > 0$ for which the family of open balls $\{B(x_n,\delta) : n \in \mathbb{N}\}$ is pairwise disjoint. Also, $\lim_{n\to\infty} t(x_n) = 0$ implies that we can assume $t(x_n) < \min\{\frac{\delta}{4},\frac{1}{n}\}$ $\forall n \in \mathbb{N}$. Let $\delta_n = \min\{\frac{\delta}{4},\frac{1}{n}\}$ $\forall n \in \mathbb{N}$. For each $n \in \mathbb{N}$, $B(x_n,\delta_n)$ is not totally bounded, thus for each n we can choose $y_n \neq x_n$ such that $d(x_n,y_n) < \delta_n$. Let $\varepsilon_n = d(x_n,y_n)$ $\forall n \in \mathbb{N}$. Define the following function.

$$f(x) = \begin{cases} 1 - \frac{d(x,x_n)}{\varepsilon_n} & : x \in B(x_n,\varepsilon_n) \text{ for some } n \in \mathbb{N} \\ 0 & : \text{otherwise} \end{cases}$$

The function f is uniformly locally Lipschitz because $\forall x \in X$, $B(x,\frac{\delta}{4})$ intersects at most one of the balls $B(x_m,\varepsilon_m)$ and f restricted to each ball $B(x_m,\varepsilon_m)$ is Lipschitz. Let $\varepsilon > 0$. Since t is uniformly continuous, $\lim_{n\to\infty} t(y_n) = 0$. Thus, the conditions $\lim_{n\to\infty} \varepsilon_n = 0$ and $\lim_{n\to\infty} t(x_n) = 0$ imply that $\exists k \in \mathbb{N}$ such that $\{x : t(x) > \varepsilon\} \cap \bigcup_{n=k+1}^{\infty} B(x_n,\varepsilon_n) = \emptyset$. Thus f restricted to $\{x : t(x) > \varepsilon\}$ is Lipschitz continuous but it is not uniformly continuous on (X,d) as $\forall n \in \mathbb{N}$, $d(x_n,y_n) < \frac{1}{n}$, but $f(x_n) - f(y_n) = 1$. We arrive at a contradiction. □

We end this section with the following example which shows that the aforementioned characterizations of metric spaces having cofinal completion do not work if we replace the functional t by ν.

Example 3.5. Consider the metric space (X,d) defined in Example 3.2. Define a function f on X as follows: $f(x) = x^2$ $\forall x \in X$. For each $n,m \in \mathbb{N}$ such that $n - \frac{1}{m} \in X$, we have $\nu(n - \frac{1}{m}) \leq \frac{1}{n}$. Clearly, f is Cauchy-regular and for each $\varepsilon > 0$, $\nu(n - \frac{1}{m}) > \varepsilon$ is possible only for finitely many $n \in \mathbb{N}$. Thus f is uniformly continuous on $\{x \in X : \nu(x) > \varepsilon\}$ as $\{x \in X : \nu(x) > \varepsilon\} \subseteq [0,M]$ for some $M \in \mathbb{N}$. But f is not uniformly continuous on (X,d) as $(n - \frac{1}{n}) \asymp (n)$ but $(f(x_n)) \not\asymp (n^2)$. Thus, even though $(\widehat{X},d)$ is cofinally complete, we have a Cauchy-regular function f on (X,d) such that $f \in CV(X,Y)$, but it is not uniformly continuous on (X,d).

3.3 Cauchy-subregular Functions

In this section, we cast light on Cauchy-subregular functions. A major part of this section is devoted to study some relations between Cauchy-subregular and CC-regular functions. We also give a condition under which a Cauchy-subregular function is Cauchy-regular or uniformly continuous.

Let us first see the following examples which show that there is no direct relation between continuous and Cauchy-subregular functions.

Example 3.6. Let $X = \{\frac{1}{n}, 0 : n \in \mathbb{N}\}$ and $Y = \{\frac{1}{n} : n \in \mathbb{N}\}$, both carrying the usual distance metric.

- Let $f : X \to \mathbb{R}$ be a function such that $f(0) = 0$, $f(\frac{1}{n}) = 1$ for odd n and $f(\frac{1}{n}) = 2$ for even n. Clearly, f is Cauchy-subregular but not continuous.
- Let $g : Y \to \mathbb{R}$ be a function such that $g\left(\frac{1}{n}\right) = n \; \forall n \in \mathbb{N}$, then g is continuous (in fact, locally Lipschitz) but not Cauchy-subregular.
- $f|_Y$ is continuous and Cauchy-subregular but not Cauchy-regular.

We know that Cauchy-subregular functions take Cauchy sequences to sequences having a Cauchy subsequence. Our first result of this section shows that Cauchy-subregular functions are precisely those functions which take Cauchy sequences to cofinally Cauchy sequences. Here we would like to recall that a cofinally Cauchy sequence in a metric space may not have any Cauchy subsequence.

Theorem 3.7. *[Gupta and Kundu (2022)] A function between two metric spaces is Cauchy-subregular if and only if it maps every Cauchy sequence to cofinally Cauchy sequence.*

Proof. It is clear that every Cauchy-subregular function sends Cauchy sequences to cofinally Cauchy sequences. For the converse part, let (x_n) be a Cauchy sequence in (X,d). Suppose $(f(x_n))$ does not have any Cauchy subsequence. Without loss of generality, let the sequence $(f(x_n))$ have all distinct elements. Since the set $A = \{f(x_n) : n \in \mathbb{N}\}$ is not totally bounded, there exists $\varepsilon > 0$ such that $\rho(f(x_n), f(x_m)) > \varepsilon \; \forall n,m \in \mathbb{N}_1$, where $\mathbb{N}_1$ is an infinite subset of $\mathbb{N}$. Now, $\{x_n : n \in \mathbb{N}_1\}$ is a Cauchy sequence in X such that its image is not cofinally Cauchy, a contradiction. □

The following proposition is an immediate corollary of the previous theorem.

Proposition 3.3. *[Gupta and Kundu (2022)] Every CC-regular function from a metric space (X,d) to any other metric space (Y,ρ) is Cauchy-subregular.*

The converse of this proposition may not be true (see Example 3.4). The following result gives necessary and sufficient conditions under which every Cauchy-subregular function is CC-regular.

Theorem 3.8. *[Gupta and Kundu (2022)] Let (X,d) be a metric space. Then the following statements are equivalent.*

- (a) *The completion $(\widehat{X},d)$ of (X,d) is cofinally complete.*
- (b) *Whenever (Y,ρ) is a metric space and $f:(X,d)\to(Y,\rho)$ is Cauchy-subregular, then f is CC-regular.*
- (c) *Whenever (Y,ρ) is a metric space and $f:(X,d)\to(Y,\rho)$ is Cauchy-subregular, then f is uniformly locally bounded.*
- (d) *Whenever (Y,ρ) is a metric space and $f:(X,d)\to(Y,\rho)$ is both continuous and Cauchy-subregular, then f is uniformly locally bounded.*
- (e) *Whenever (Y,ρ) is a metric space and $f:(X,d)\to(Y,\rho)$ is both locally Lipschitz and Cauchy-subregular, then f is uniformly locally bounded.*
- (f) *If $f:(X,d)\to\mathbb{R}$ is both locally Lipschitz and Cauchy-subregular, then f is uniformly locally bounded.*

Proof. The implications $(c)\Rightarrow(d)\Rightarrow(e)\Rightarrow(f)$ are all immediate.

$(a)\Rightarrow(b)$: Let $f:(X,d)\to(Y,\rho)$ be a Cauchy-subregular function. To see that f is CC-regular, let (x_n) be a cofinally Cauchy sequence in X. Since $(\widehat{X},d)$ is cofinally complete, (x_n) has a Cauchy subsequence say (x_{n_k}). Hence $(f(x_{n_k}))$ has a Cauchy subsequence which implies that $(f(x_n))$ is cofinally Cauchy. Thus f is CC-regular.

$(b)\Rightarrow(c)$: This implication follows from Theorem 2.9.

$(f)\Rightarrow(a)$: Suppose $(\widehat{X},d)$ is not cofinally complete. Thus there exists a cofinally Cauchy sequence $(\hat{x_n})$ in $\widehat{X}$ which has no Cauchy subsequence. Since X is dense in $\widehat{X}$, for each $n\in\mathbb{N}$, there exists $x_n\in X$ such that $d(x,\hat{x})<\frac{1}{n}$. Thus (x_n) is a cofinally Cauchy sequence in (X,d) without any Cauchy subsequence. By passing to a subsequence, we can assume that (x_n) consists of distinct points. Thus, for each n, there exists $0<\delta_n<\frac{1}{n}$ such that the family of open balls $\{B(x_n,\delta_n):n\in\mathbb{N}\}$ is pairwise disjoint. Define the following function.

$$f(x)=\begin{cases} n-\frac{n}{\delta_n}d(x,x_n) & : x\in B(x_n,\delta_n) \text{ for some } n\in\mathbb{N} \\ 0 & : \text{otherwise} \end{cases}$$

It can be verified that f restricted to each $B(x_n,\delta_n)$ is Lipschitz and thus is locally Lipschitz. We now claim that f is Cauchy-subregular. Let (z_n) be any Cauchy sequence in (X,d) and let $B=\{z_n:n\in\mathbb{N}\}$. Since (x_n) has no Cauchy subsequence, the situation: $B\cap B(x_n,\delta_n)\neq\emptyset$ is only possible for finitely many

$n \in \mathbb{N}$. Thus it follows that there exists $M_B > 0$ such that $f(B) \subseteq [0, M_B]$. Hence $f(B)$ is totally bounded and consequently $(f(z_n))$ has a Cauchy subsequence. It remains to show that f is not uniformly locally bounded. On the contrary, suppose there exists $\delta > 0$ such that $f(B(x,\delta))$ is bounded for each $n \in \mathbb{N}$. Since (x_n) is cofinally Cauchy, there exists an infinite subset $\mathbb{N}_o$ of $\mathbb{N}$ such that $0 < d(x_n, x_m) < \delta\ \forall n, m \in \mathbb{N}_o$, but their image is not bounded, a contradiction. $\square$

In view of Theorem 3.7, let us study some more sequential characterizations of Cauchy-subregular functions.

Theorem 3.9. *Let $f : (X,d) \to (Y,\rho)$ be a function between two metric spaces. Then the following assertions are equivalent.*

- (a) *f is Cauchy-subregular.*
- (b) *Every sequence (x_n) has a subsequence on which f is uniformly continuous.*
- (c) *Every sequence (x_n) has a subsequence on which f is Cauchy-regular.*
- (d) *Every sequence (x_n) has a subsequence on which f is CC-regular.*

Proof. $(a) \Rightarrow (b)$: Let (x_n) be any sequence in X. If (x_n) has no Cauchy subsequence, then there exists a $\delta > 0$ such that, by passing to a subsequence, $d(x_n, x_m) > \delta$ for all $n, m \in \mathbb{N}$. Thus f is uniformly continuous on (x_n). If (x_n) has Cauchy subsequence, then without loss of generality, we can assume that (x_n) is Cauchy. Let us also assume that all the elements of (x_n) are distinct, because otherwise, f will be uniformly continuous on that constant subsequence. Now, since f is Cauchy-subregular, there exists a Cauchy subsequence $(f(x_{n_k}))$ of $(f(x_n))$ in (Y,ρ). If the sequence (x_{n_k}) converges to some point $x \in X$, and if $x_{n_k} = x$ for some $k \in \mathbb{N}$, then remove the point x_{n_k} from the sequence. We now claim that f is uniformly continuous on the set $A = \{x_{n_k} : k \in \mathbb{N}\}$. Let $\varepsilon > 0$. Thus there exists $k_o \in \mathbb{N}$ such that $\rho(f(x_{n_l}), f(x_{n_m})) < \varepsilon$ for all $l, m \geqslant k_o$. For each $1 \leq i \leq k_o$, there exists $\varepsilon_i > 0$ such that $d(x_{n_i}, x_{n_t}) > 2\varepsilon_i$ for all $t \neq i$. Now $\delta = \min\{d(x_{n_l}, x_{n_m}), \varepsilon_i : 1 \leq l \leq k_o,\ 1 \leq m \leq k_o, 1 \leq i \leq k_o\}$ will do the job.

$(b) \Rightarrow (c)$: This is immediate.

$(c) \Rightarrow (d)$: Let (x_n) be a sequence in X. If $A = \{x_n : n \in \mathbb{N}\}$ is not totally bounded, then there exist a subsequence (x_{n_k}) of (x_n) and $\varepsilon > 0$ such that $d(x_{n_k}, x_{n_t}) > \varepsilon\ \forall k, t \in \mathbb{N}$. Thus f is CC-regular on (x_{n_k}). If A is totally bounded, we can assume that (x_n) is Cauchy. Thus there exists a subsequence (x_{n_k}) of (x_n) such that $(f(x_{n_k}))$ is Cauchy. Hence f is CC-regular on (x_{n_k}).

$(d) \Rightarrow (a)$: Let (x_n) be a Cauchy sequence in (X,d). Thus there exists a subsequence (x_{n_k}) of (x_n) such that f is CC-regular on (x_{n_k}). Since (x_{n_k}) is

Cauchy, $(f(x_{n_k}))$ is cofinally Cauchy, which further implies that $(f(x_n))$ is cofinally Cauchy. Hence the result follows from Theorem 3.7. □

Remark 3.4. The equivalence of (a) and (b) of Theorem 3.9 was proved in [Beer and Levi (2009a)], and the equivalence of (a), (c) and (d) was proved in [Gupta and Kundu (2022)].

In Theorem 3.7, we have noticed that a Cauchy-subregular function takes every Cauchy sequence to cofinally Cauchy sequence. Now we look for the condition under which a Cauchy-subregular function takes cofinally Cauchy sequences to sequences having a Cauchy subsequence.

Theorem 3.10. *[Gupta and Kundu (2022)] Let (X,d) be a metric space. Then the following conditions are equivalent.*

- (a) *The completion $(\widehat{X},d)$ of (X,d) is cofinally complete.*
- (b) *Whenever (Y,ρ) is a metric space and $f:(X,d)\to(Y,\rho)$ is Cauchy-subregular, then f maps cofinally Cauchy sequences to sequences having a Cauchy subsequence.*

Proof. $(a)\Rightarrow(b)$: Let $f:(X,d)\to(Y,\rho)$ be a Cauchy-subregular function and let (x_n) be a cofinally Cauchy sequence in (X,d). Since $(\widehat{X},d)$ is cofinally complete, (x_n) has a Cauchy subsequence. As f is Cauchy-subregular, $(f(x_n))$ has a Cauchy subsequence.

$(b)\Rightarrow(a)$: Suppose $(\widehat{X},d)$ is not cofinally complete. Then there exists a cofinally Cauchy sequence (x_n) of distinct points in (X,d) such that it has no Cauchy subsequence. Define the following function.

$$f(x)=\begin{cases} n: x=x_n \text{ for some } n\in\mathbb{N} \\ 0: \text{otherwise} \end{cases}$$

Since (x_n) has no Cauchy subsequence, f is Cauchy-subregular but clearly $(f(x_n))$ has no Cauchy subsequence, a contradiction. □

Evidently, every Cauchy-regular function is Cauchy-subregular but the converse is not true. To find out: when does a Cauchy-subregular function satisfy Cauchy-regularity, let us first recall what is meant by a bornology.

Definitions 3.4. A family $\mathbf{B}$ of subsets of a metric space (X,d) is said to be *bornology* if (i) $\mathbf{B}$ forms a cover of X (ii) $\mathbf{B}$ is hereditary, that is, if $A\in\mathbf{B}$ and $B\subseteq A$, then $B\in\mathbf{B}$ (iii) $\mathbf{B}$ is closed under finite unions.

A family of subsets $\mathscr{B}_o$ is said to be a base for $\mathbf{B}$ if $\forall\, B\in\mathbf{B}$, $\exists B_o\in\mathscr{B}_o$ such that $B\subseteq B_o$. If each member of a base is a closed subset of X, then the base is called a *closed base*.

As an example, note that the smallest bornology on X is the family of finite subsets of X and the largest one is the power set $\mathbb{P}(X)$ of X. We refer the interested readers to [Beer and Levi (2009b)] and the references therein for more information on bornologies. Note that the family of totally bounded subsets of X also forms a bornology. So by Proposition 1.9, a function $f : (X,d) \to (Y,\rho)$ is Cauchy-subregular if and only if it maps members of the bornology of totally bounded subsets of (X,d) to members of the corresponding bornology of (Y,ρ).

Proposition 3.4. *Let $f : (X,d) \to (Y,\rho)$ be a function between two metric spaces. Then the family of subsets on which f is Cauchy-subregular forms a bornology $\mathbf{B_s}$. Moreover, if f is continuous, then $\mathbf{B_s}$ has a closed base.*

Proof. We will prove that if f is Cauchy-subregular on A and B, where A and B are subsets of X, then f is Cauchy-subregular on $A \cup B$. Let (x_n) be a Cauchy sequence in $A \cup B$. Thus there exists a subsequence (x_{n_k}) of (x_n) which lies either in A or in B. Since f is Cauchy-subregular on both A and B, there exists a Cauchy subsequence of $(f(x_n))$ in (Y,ρ).

Next, let f be continuous. Let f be Cauchy-subregular on a subset A of X. To prove that f is Cauchy-subregular on the closure of A, say $\overline{A}$, let (x_n) be a Cauchy sequence in $\overline{A}$. Since f is continuous, for each $n \in \mathbb{N}$, there exists $y_n \in A$ such that $d(x_n, y_n) < \frac{1}{n}$ and $\rho(f(x_n), f(y_n)) < \frac{1}{n}$. So $(y_n) \subseteq A$ is a Cauchy sequence. Thus $(f(y_n))$ has a Cauchy subsequence, which further implies that $(f(x_n))$ has a Cauchy subsequence too. Hence we are done. □

We leave it to the interested readers to produce a (continuous) function on $\{\frac{1}{n} : n \in \mathbb{N}\}$ such that the subsets on which the function is Cauchy regular are not closed under finite unions. Now we add an extra condition on Cauchy-subregular functions to make it Cauchy-regular. For proving the result, we need the following lemma from [Beer (1993)].

Lemma 3.3. *[Beer (1993), p. 92] Efremovic Lemma: Let (X,d) be a metric space. If (x_n) and (y_n) are two sequences in X such that for some $\varepsilon > 0$, $d(x_n, y_n) > \varepsilon\ \forall n \in \mathbb{N}$, then there exists an infinite subset $\mathbb{N}_1$ of $\mathbb{N}$ such that $d(x_k, y_l) \geqslant \frac{\varepsilon}{4}\ \forall k, l \in \mathbb{N}_1$.*

Theorem 3.11. *[Gupta and Kundu (2022)] Let $f : (X,d) \to (Y,\rho)$ be a function between two metric spaces. Then the following statements are equivalent.*

(*a*) *f is Cauchy-regular.*
(*b*) *f is Cauchy-subregular and the family of subsets on which f is Cauchy-regular is closed under finite unions.*

(c) *f is Cauchy-subregular and the family of subsets on which f is Cauchy-regular forms a bornology.*

Proof. The implications $(a) \Rightarrow (b) \Rightarrow (c)$ are immediate.

$(c) \Rightarrow (a)$: Suppose f is not Cauchy-regular. Thus there exists a Cauchy sequence (x_n) in (X,d) such that $(f(x_n))$ is not Cauchy. It follows that f is not uniformly continuous on the set $A = \{x_n : n \in \mathbb{N}\}$. So for some $\varepsilon > 0$, there exist sequences (y_n) and (z_n) in A such that $d(y_n, z_n) < \frac{1}{n}$ but $\rho(f(y_n), f(z_n)) > \varepsilon \; \forall n \in \mathbb{N}$. By the Efremovic Lemma, we can assume that $\{y_n : n \in \mathbb{N}\} \cap \{z_n : n \in \mathbb{N}\} = \emptyset$. Since $(y_n) \subseteq A$, which is totally bounded, by passing to a subsequence we can assume that (y_n) is Cauchy. Hence the corresponding sequence (z_n) is also Cauchy. Since f is Cauchy-subregular, there exists an infinite subset $\mathbb{N}_1$ of $\mathbb{N}$ such that $(f(y_n))_{n \in \mathbb{N}_1}$ is Cauchy in Y and similarly, there exists an infinite subset $\mathbb{N}_2$ of $\mathbb{N}_1$ such that $(f(z_n))_{n \in \mathbb{N}_2}$ is Cauchy in Y. Thus f is Cauchy-regular on $\{y_n : n \in \mathbb{N}_2\}$ and $\{z_n : n \in \mathbb{N}_2\}$ but not on their union because if we enumerate the elements of $\mathbb{N}_2$ in the increasing order say $n_1, n_2, n_3, \ldots$, then the sequence $y_{n_1}, z_{n_1}, y_{n_2}, z_{n_2}, \ldots$ is Cauchy but its image is not. We get a contradiction. Hence f is Cauchy-regular. □

Note that if we define a function f from $X = \{\frac{1}{n} : n \in \mathbb{N}\}$ to $\mathbb{R}$ such that $f(\frac{1}{n}) = n$ for all $n \in \mathbb{N}$, then the family of subsets on which f is Cauchy-regular forms a bornology but f is not Cauchy-regular. Hence the Cauchy-subregularity of f in Theorem 3.11 is required. We have the following analogous characterization for the class of uniformly continuous functions. Its proof is similar to that of Theorem 3.11.

Theorem 3.12. *[Gupta and Kundu (2022)] Let $f : (X,d) \to (Y,\rho)$ be a function between two metric spaces. Then the following assertions are equivalent.*

(a) *f is uniformly continuous.*
(b) *f is Cauchy-subregular and the family of subsets on which f is uniformly continuous is closed under finite unions.*
(c) *f is Cauchy-subregular and the family of subsets on which f is uniformly continuous forms a bornology.*

We would also like to make the following observation about CC-regular functions.

Theorem 3.13. *Let $f : (X,d) \to (Y,\rho)$ be a function between two metric spaces. The family of subsets on which f is CC-regular forms a bornology.*

Proof. Let A and B be two subsets of X. We only need to prove that if f is CC-regular on A and B, then f is CC-regular on $A \cup B$. Let (x_n) be a cofinally Cauchy sequence in $A \cup B$. By Proposition 1.1, there exists a pairwise disjoint family $\{\mathbb{M}_j : j \in \mathbb{N}\}$ of infinite subsets of $\mathbb{N}$ such that if $i, l \in \mathbb{M}_j$ then $d(x_i, x_l) < \frac{1}{j}$. For each $j \in \mathbb{N}$, let $A_j = \{x_i : i \in M_j\}$. Thus each A_j contains infinitely many elements atleast from A or B or both. Thus atleast one of A or B would contain infinitely elements from infintely many $\{A_j : j \in \mathbb{N}\}$. Hence (x_n) has a cofinally Cauchy subsequence that lies in either A or B. Since f is CC-regular on A and B, $(f(x_n))$ is cofinally Cauchy. Hence f is CC-regular on $A \cup B$. □

In view of the observation that the family of subsets on which a function is Cauchy-regular need not form a bornology, let us note a modified characterization of complete metric spaces. The proof is left to the reader.

Theorem 3.14. *[Gupta and Kundu (2022)] Let (X, d) be a metric space. Then the following statements are equivalent.*

- (a) *(X, d) is complete.*
- (b) *Whenever (Y, ρ) is a metric space and $f : (X, d) \to (Y, \rho)$ is continuous, then the family of subsets on which f is Cauchy-regular forms a bornology.*
- (c) *If $f : (X, d) \to \mathbb{R}$ is continuous, then the family of subsets on which f is Cauchy-regular forms a bornology.*

3.4 Some More Characterizations

In this section, we give some characterizations of metric spaces having cofinal completion in terms of variants of the finite intersection property and Cantor-type results.

Analogous to the concept of almost nowhere locally compact sets considered in Chapter 2, we define a non-empty subset A of X to be almost nowhere locally totally bounded if $\forall\, \varepsilon > 0$, the set $\{a \in A : t(a) \geq \varepsilon\}$ is totally bounded. We denote the set of closed almost nowhere locally totally bounded subsets by $AT(X)$ and the set of complete almost nowhere locally totally bounded subsets by $CAT(X)$. Recall that a subset A of a metric space (X, d) is said to have *countable character* if for every open set G containing A, there exists $r > 0$ such that $A \subseteq B(A, r) \subseteq G$. It is well known that in a metric space, every compact subset has a countable character. In the next result we show that a metric space (X, d) has a cofinal completion if and only if every $A \in CAT(X)$ has a countable character.

Note that in Example 3.2, if we take the set $A = \{x_n : n \in \mathbb{N}\}$, then $A \in AC(X)$ but $A \notin AT(X)$.

Theorem 3.15. *[Gupta and Kundu (2021b)] Let (X,d) be a metric space. Then the following statements are equivalent:*

(a) *The completion $(\widehat{X},d)$ of (X,d) is cofinally complete.*
(b) *If B and C are non-empty disjoint closed subsets of X such that $C \in CAT(X)$, then $D(B,C) > 0$.*
(c) *If $A \in CAT(X)$, then A has countable character.*
(d) *If B and C are non-empty disjoint closed subsets of X such that B is complete and $C \in AT(X)$, then $D(B,C) > 0$.*
(e) *If B and C are non-empty disjoint closed subsets of X such that B is complete and $C \in CAT(X)$, then $D(B,C) > 0$.*
(f) *Let (A_n) be a sequence of closed sets in X such that for some $k \in \mathbb{N}$, A_k is complete in X and for some $t \in \mathbb{N}$ $(t \neq k)$, $A_t \in AT(X)$. If $\bigcap_{n=1}^{\infty} A_n = \emptyset$, then there exists $r > 0$ such that $\bigcap_{n=1}^{\infty} B(A_n, r) = \emptyset$.*
(g) *Let (A_n) be a sequence of complete sets in X such that for some $k \in \mathbb{N}$, $A_k \in AT(X)$. If $\bigcap_{n=1}^{\infty} A_n = \emptyset$, then there exists $r > 0$ such that $\bigcap_{n=1}^{\infty} B(A_n, r) = \emptyset$.*
(h) *Let (A_n) be a sequence of closed sets in X such that for some $k \in \mathbb{N}$, $A_k \in CAT(X)$. If $\bigcap_{n=1}^{\infty} A_n = \emptyset$, then there exists $r > 0$ such that $\bigcap_{n=1}^{\infty} B(A_n, r) = \emptyset$.*

Proof. $(a) \Rightarrow (b)$: Let $(\widehat{X},d)$ be cofinally complete. Let B be a closed subset of X and $C \in CAT(X)$ such that $B \cap C = \emptyset$. We first claim that $C \in AC(\widehat{X})$. Let $\varepsilon > 0$. Being a closed subset of a complete set, the totally bounded set $A = \{x \in C : t(x) \geqslant \varepsilon\}$ is complete. Thus the set $\{x \in C : \hat{\nu}(x) \geqslant \varepsilon\} = \{x \in C : t(x) \geqslant \varepsilon\}$ is compact which implies $C \in AC(\widehat{X})$. It is easy to see that $\widehat{B} \in CL(\widehat{X})$ and $C \in AC(\widehat{X})$ are disjoint in $(\widehat{X},d)$. Since $(\widehat{X},d)$ is cofinally complete, by Theorem 2.7, $D(\widehat{B},C) > 0$ and hence $D(B,C) > 0$.

$(b) \Rightarrow (c)$: Let $A \in CAT(X)$ and $A \subseteq G$ for some open set G in X. Since G^c is closed and $A \cap G^c = \emptyset$. By hypothesis, $D(A,G^c) > r > 0$. Hence $A \subseteq B(A,r) \subseteq G$.

$(c) \Rightarrow (d)$: Let B and C be disjoint subsets of X such that B is complete and $C \in AT(X)$. Suppose $D(B,C) = 0$. Then for each $n \in \mathbb{N}$, we can find $x_n \in B$ and $y_n \in C$ such that $d(x_n,y_n) < \frac{1}{n}$. If (x_n) has a Cauchy subsequence, it will have a cluster point z in B and thus $z \in C$ as C is closed and z would be a cluster point of (y_n) too. We get a contradiction. Hence neither (x_n) nor (y_n) has a Cauchy subsequence. Consider the disjoint sets $B_1 = \{x_n : n \in \mathbb{N}\}$ and $C_1 = \{y_n : n \in \mathbb{N}\}$. Clearly, $C_1 \in CAT(X)$, B_1 is closed and $C_1 \subseteq B_1^c$, but there does not exist any $r > 0$ such that $C_1 \subseteq B(C_1,r) \subseteq B_1^c$.

$(d) \Rightarrow (e)$: This is immediate.

$(e) \Rightarrow (a)$: To prove that $(\widehat{X},d)$ is cofinally complete, let B be a closed subset of $\widehat{X}$ and $C \in AC(\widehat{X})$ such that B and C are disjoint. If we show that $D(B,C) > 0$, then by Theorem 2.7, we will be done. Suppose this is not true. Thus there exists sequences $(x_n) \subseteq B$ and $(y_n) \subseteq C$ such that for each $n \in \mathbb{N}$, $d(x_n,y_n) < \frac{1}{n}$. Clearly, the sequences (x_n) and (y_n) do not cluster and thus for each $n \in \mathbb{N}$, $\exists\, \delta'_n > 0$ and $\delta''_n > 0$ such that the family of open balls $\{B(x_n,\delta'_n), B(y_n,\delta''_n) : n \in \mathbb{N}\}$ is pairwise disjoint. Choose sequences (x'_n) and (y'_n) in X such that $d(x_n,x'_n) < \frac{\delta'_n}{2}$ and $d(y_n,y'_n) < \frac{\delta''_n}{2}$. Let $Y = \{x'_n : n \in \mathbb{N}\}$ and $Z = \{y'_n : n \in \mathbb{N}\}$. It can be verified that Y and Z are complete subset of X. Further to see that $Z \in CAT(X)$, let $\varepsilon > 0$. Choose $n_o > \frac{2}{\varepsilon}$ and thus $\forall n \geqslant n_o$, if $t(y'_n) \geqslant \varepsilon$, then $\hat{\nu}(y_n) \geqslant \frac{\varepsilon}{2}$ as $B(y_n,\frac{\varepsilon}{2}) \subseteq B(y'_n,\varepsilon)$. Since (y_n) has no cluster point and $C \in AC(\widehat{X})$, the situation $\hat{\nu}(y_n) \geqslant \frac{\varepsilon}{2}$ is only possible for finitely many elements from $\{y_n : n \in \mathbb{N}\}$. Hence $t(y'_n) \geqslant \varepsilon$ is possible for finitely many elements from $\{y'_n : n \in \mathbb{N}\}$. Thus we have a complete subset Y and $Z \in CAT(X)$ such that Y and Z are disjoint and $D(Y,Z) = 0$, which is a contradiction.

$(a) \Rightarrow (f)$: Let (A_n) be a sequence of closed sets in X such that for some $k \in \mathbb{N}$, A_k is complete in X and for some $t \in \mathbb{N}$ $(t \neq k)$, A_t is almost nowhere locally totally bounded and $\bigcap_{n=1}^{\infty} A_n = \emptyset$. Suppose $\bigcap_{n=1}^{\infty} B(A_n, 1/m) \neq \emptyset\ \forall m \in \mathbb{N}$. Let $x_m \in \bigcap_{n=1}^{\infty} B(A_n, 1/m)\ \forall m \in \mathbb{N}$. In particular, $x_m \in B(A_k, 1/m)\ \forall m \in \mathbb{N}$. Hence for each $m \in \mathbb{N}$, we can find $y_m \in A_k$ such that $d(x_m,y_m) < 1/m$.

Now if (x_m) has a Cauchy subsequence, then so has the sequence (y_m), say (y_{m_k}). Since $(y_m) \in A_k$ and A_k is complete, the sequence (y_{m_k}) converges to some point $x \in A_k$. Hence the sequence (x_{m_k}) also converges to the point x. Note that for any $t \in \mathbb{N}$, $d(x,A_t) \leq d(x,x_{m_k}) + d(x_{m_k},A_t)$. Since $x_{m_k} \in B(A_t, 1/m_k)$, $d(x_{m_k},A_t) < 1/m_k < 1/k$ and consequently $\lim_{k\to\infty} d(x_k,A_t) = 0$. Hence $d(x,A_t) = 0$. Since A_t is closed, $x \in A_t$ and this is true for all $t \in \mathbb{N}$. Thus, $x \in \bigcap_{n=1}^{\infty} A_n$. We get a contradiction. Hence the sequence (x_m) cannot have any Cauchy subsequence.

Now, as we chose $(y_m) \subseteq A_k$, choose a new sequence $(z_m) \subseteq A_t$. Since A_t is almost nowhere locally totally bounded and (z_m) does not have any Cauchy subsequence, we can find a subsequence (z_{m_k}) of (z_m) such that $\lim_{k\to\infty} t(z_{m_k}) = 0$. Since the sequence (z_{m_k}) does not have any Cauchy subsequence, by Theorem 3.2 we get a contradiction.

In a manner similar to the proof of $(a) \Rightarrow (f)$, we can prove $(a) \Rightarrow (h)$. The implications $(f) \Rightarrow (g)$, $(h) \Rightarrow (g) \Rightarrow (e)$ are all immediate. □

Let us see an example where $(\widehat{X},d)$ is not cofinally complete and the condition (e) of the previous theorem fails.

Example 3.7. Consider the metric space (X,d) defined in Example 3.1. Let $A = \{e_n + \frac{1}{n}e_1 : n \in \mathbb{N}\}$ and $B = \{e_n + \frac{1}{n}e_2 : n \in \mathbb{N}\}$. Clearly, both the sets A and B belong to $CAT(X)$ as both the sets are complete and for all $\varepsilon > 0$, the set $\{z \in A \cup B : t(a) \geqslant \varepsilon\}$ is finite. Here, A and B are disjoint but $D(A,B) = 0$.

We end this chapter by giving some Cantor-type characterizations which are suggested by Theorem 2.7. Let us first study the following relevant definitions.

Definition 3.5. Let (X,d) be a metric space and A be a non-empty subset of X. Then we define $\overline{t}(A) = \sup\{t(x) : x \in A\}$ and $\underline{t}(A) = \inf\{t(x) : x \in A\}$.

Let B and C be disjoint subsets of (X,d). Then B and C are said to be asymptotic if $\forall\, \varepsilon > 0$, $\exists\, b \in B$ and $c \in C$ with $d(b,c) < \varepsilon$.

We skip the proof of the following set of characterizations as they can be proved by using Theorem 2.7.

Theorem 3.16. *[Gupta and Kundu (2021b)] Let (X,d) be a metric space. Then the following statements are equivalent:*

- (a) *The completion $(\widehat{X},d)$ of (X,d) is cofinally complete.*
- (b) *If F_1 and F_2 are disjoint asymptotic subsets of X such that F_1 is closed and F_2 is complete, then there exists a $\delta > 0$ such that $F_1 \cap \{x \in X : t(x) > \delta\}$ and $F_2 \cap \{x \in X : t(x) > \delta\}$ are asymptotic.*
- (c) *If F_1 and F_2 are disjoint asymptotic complete subsets of X, then there exists a $\delta > 0$ such that $F_1 \cap \{x \in X : t(x) > \delta\}$ and $F_2 \cap \{x \in X : t(x) > \delta\}$ are asymptotic.*
- (d) *If (F_n) is a decreasing sequence of non-empty complete subsets of X with $\lim_{n\to\infty} \underline{t}(F_n) = 0$, then $\bigcap\{F_n : n \in \mathbb{N}\}$ is non-empty.*
- (e) *If (F_n) is a decreasing sequence of non-empty closed subsets of X such that for some $m \in \mathbb{N}$, F_m is complete and $\lim_{n\to\infty} \underline{t}(F_n) = 0$, then $\bigcap\{F_n : n \in \mathbb{N}\}$ is non-empty.*
- (f) *If (F_n) is a decreasing sequence of non-empty closed subsets of X such that for some $m \in \mathbb{N}$, F_m is complete and $\lim_{n\to\infty} \overline{t}(F_n) = 0$, then $\bigcap\{F_n : n \in \mathbb{N}\}$ is non-empty.*
- (g) *If (F_n) is a decreasing sequence of non-empty complete subsets of X with $\lim_{n\to\infty} \overline{t}(F_n) = 0$, then $\bigcap\{F_n : n \in \mathbb{N}\}$ is non-empty.*

For additional reading, one may also refer to [Beer *et al.* (2011)].

Exercises

Exercise 3.1

Prove Proposition 3.1 and Theorem 3.14.

Exercise 3.2

Give example of a complete and locally compact subset of (X,d) which is not uniformly locally compact in its relative topology.

Exercise 3.3

Give examples of:

(a) a t-bounded function which is not bounded.
(b) uniformly locally Lipschitz function which is not t-bounded.
(c) Cauchy-subregular function which is not uniformly locally bounded.

Exercise 3.4

[Beer and Levi (2009a)] Let $f : (X,d) \to (Y,\rho)$ be a function between two metric spaces. Then show the equivalence of the following statements:

(a) f is Cauchy-regular.
(b) f is strongly uniformly continuous on the totally bounded subsets of (X,d).
(c) f is uniformly continuous on the totally bounded subsets of (X,d).

Exercise 3.5

Show that the following collections form bornologies:

(a) the collection of all nowhere dense subsets of a metric space (X,d) where X has no isolated points.
(b) the set of finitely chainable subsets of a metric space (X,d).
(c) intersection of a non-empty family of bornologies.
(d) (Beer and Levi [Beer and Levi (2009a)]) $\mathscr{B}_f = \{B \subseteq X : B \neq \emptyset,\ f \text{ is strongly uniformly continuous on } B\}$ where f is a continuous function from (X,d) to (Y,ρ). Moreover, verify that $\mathscr{B}_f$ is a bornology if and only if f is continuous.

Exercise 3.6

[Beer and Levi (2009a)] Let $f:(X,d)\to(Y,\rho)$ be a function and (x_n) be a Cauchy sequence in X with distinct terms. Then prove that f restricted to $\{x_n : n\in\mathbb{N}\}$ is uniformly continuous if and only if $(f(x_n))$ is a Cauchy sequence in Y.

Exercise 3.7

[Beer and Levi (2009a)] Let $f:(X,d)\to(Y,\rho)$ be a Cauchy-subregular function. Prove that f is continuous if and only if $\mathscr{B}_f$ has a closed base.

Exercise 3.8

[Gupta and Kundu (2022)] Let (X,d) be a metric space. Prove that the following are equivalent.

(a) (X,d) is complete.
(b) Whenever (Y,ρ) is a metric space and $f:(X,d)\to(Y,\rho)$ is continuous, then f is Cauchy-subregular.
(c) Whenever (Y,ρ) is a metric space and $f:(X,d)\to(Y,\rho)$ is locally Lipschitz, then f is Cauchy-subregular.
(d) Whenever (Y,ρ) is a metric space and $f:(X,d)\to(Y,\rho)$ is both continuous and Cauchy-subregular, then f is Cauchy-regular.
(e) Whenever (Y,ρ) is a metric space and $f:(X,d)\to(Y,\rho)$ is both locally Lipschitz and Cauchy-subregular, then f is Cauchy-regular.
(f) Whenever $(Z,\|.\|)$ is a Banach space and $f:(X,d)\to(Z,\|.\|)$ is both continuous and Cauchy-subregular, then f can be uniformly approximated by Cauchy-Lipschitz functions.
(g) Whenever $(Z,\|.\|)$ is a Banach space and $f:(X,d)\to(Z,\|.\|)$ is both locally Lipschitz and Cauchy-subregular, then f can be uniformly approximated by Cauchy-Lipschitz functions.

Exercise 3.9

[Jain and Kundu (2007)] Let (X,d) be a metric space and $(\widehat{X},d)$ be its completion. Then show that the following statements are equivalent:

(a) For every metric space (Y,ρ), every Cauchy-regular function $f:(X,d)\to(Y,\rho)$ is uniformly continuous.
(b) Every real-valued Cauchy-regular function on (X,d) is uniformly continuous.
(c) The metric space $(\widehat{X},d)$ is a UC space.

(d) For any two disjoint sets A_1 and A_2 in X with A_1 complete and A_2 closed, $D(A_1, A_2) > 0$.

(e) Every real-valued bounded Cauchy-regular function on (X, d) is uniformly continuous.

(f) Every pseudo-Cauchy sequence with distinct terms in (X, d) has a Cauchy subsequence.

Chapter 4

Cofinal Completeness vis-$\grave{a}$-vis Hyperspaces

As the name suggests, this chapter emphasizes on the characterizations of cofinally complete metric spaces in terms of some hyperspace and function space topologies. For a metric space (X,d), we consider some hyperspace topologies on the set $AC(X)$ of almost nowhere locally compact sets, which is a subspace of the space of closed subsets in X, and characterize cofinally complete metric spaces in relations to Hausdorff metric topology, proximal topology, Vietoris topology and locally finite topology on $AC(X)$. Furthermore, the class of metric spaces, for which the corresponding space $AC(X)$ equipped with the Hausdorff metric topology is cofinally complete, are discussed.

4.1 Some Hyperspace Topologies

Let us define some hyperspace topologies in terms of which we will characterize cofinally complete metric spaces.

Definition 4.1. Let (X,τ) be a topological space. The collection of all non-empty closed subsets of X is called the hyperspace of X and is denoted by $CL(X)$.

Definition 4.2. For a metric space (X,d), the *Hausdorff metric* H_d on $CL(X)$ is defined by

$$H_d(A_1,A_2) = \max\left\{\sup_{x\in A_1} d(x,A_2),\ \sup_{y\in A_2} d(y,A_1)\right\} \quad \text{for all } A_1,A_2 \in CL(X),$$

equivalently,

$$H_d(A_1,A_2) = \inf\{\varepsilon > 0 : A_1 \subset B(A_2,\varepsilon),\ A_2 \subset B(A_1,\varepsilon)\} \quad \text{for all } A_1,A_2 \in CL(X)$$

We denote the topology generated by the above defined metric H_d by τ_{H_d} and call it the Hausdorff metric topology. Note that the metric H_d may assume the

value $+\infty$. Also, if we have two equivalent metrics on a set X, they need not determine equivalent Hausdorff metrics on $CL(X)$. In fact, two metrics on a set X generate the same Hausdorff metric topology on $CL(X)$ if and only if they are uniformly equivalent [Beer (1993), Theorem 3.3.2]. Thus if (X,d) is a metric space, then the metric $D = \min\{1,d\}$, being uniformly equivalent to d, generates the same hyperspace topology. Hence in order to deal with $(CL(X), H_d)$, we can only consider bounded metrics on X. Let us now consider the following subsets of $CL(X)$ for a metric space (X,d).
$W^+ = \{B \in CL(X) : B \subseteq W\}$,
$W^{++} = \{B \in CL(X) : \text{there exists } \delta > 0 \text{ such that } B(B,\delta) \subseteq W\}$,
$\mathscr{A}_1^- = \{B \in CL(X) : \text{for all } A \in \mathscr{A}_1,\ B \cap A \neq \emptyset\}$,
$\mathscr{A}_2^- = \{B \in CL(X) : \text{for all } A \in \mathscr{A}_2,\ B \cap A \neq \emptyset\}$,
where W is an open subset of X, $\mathscr{A}_1$ is a finite family of open sets in X and $\mathscr{A}_2$ is a locally finite family of open sets in X.

Keeping these notations in mind, let us study the following definitions.

Definition 4.3. Let (X,d) be a metric space. Then a topology on $CL(X)$ is said to be:

(a) *upper Vietoris topology* (τ_{V^+}), if it is generated by the basic open sets of the form W^+.
(b) *Vietoris topology* (τ_V), if it is generated by the subbasic open sets of the form W^+ and $\mathscr{A}_1^-$.
(c) *locally finite topology* (τ_{lf}), if it is generated by the subbasic open sets of the form W^+ and $\mathscr{A}_2^-$.
(d) *proximal topology* (τ_{δ_d}), if it is generated by the subbasic open sets of the form W^{++} and $\mathscr{A}_1^-$.

Some basic facts about the Vietoris topology can be found in [Michael (1951)]. For a metric space (X,d), it is an interesting exercise to check that on $CL(X)$, $\tau_{V^+} \subset \tau_V \subseteq \tau_{lf}$, $\tau_{\delta_d} \subseteq \tau_{H_d} \subseteq \tau_{lf}$ and $\tau_{\delta_d} \subseteq \tau_V$ [Beer (1993)]. In this chapter, we also consider some hyperspace topologies on the set of continuous functions from a metric space (X,d) to another metric space (Y,ρ) and this set is denoted by $C(X,Y)$.

Definition 4.4. Let $C(X,Y)$ be the collection of all continuous functions from a metric space (X,d) to another metric space (Y,ρ). For $f \in C(X,Y)$ and $\varepsilon > 0$, we define

$$\langle f, X, \varepsilon \rangle = \{g \in C(X,Y) : \rho(f(x), g(x)) < \varepsilon \text{ for all } x \in X\}$$

Let τ_{uc} be the collection of subsets of $C(X,Y)$ such that $W \in \tau_{uc}$ if for all $f \in W$, there exists $\varepsilon > 0$ for which $\langle f, X, \varepsilon \rangle \subseteq W$. Then τ_{uc} is called the topology of uniform convergence.

Note that τ_{uc} is metrizable by the (infinite valued) uniform metric d_{uc} on $C(X,Y)$ defined by $d_{uc}(f,g) = \sup\{\rho(f(x), g(x)) : x \in X\}$. It is noteworthy that every continuous function $f : (X,d) \to (Y,\rho)$ can be considered as a closed subset of $X \times Y$ endowed with the product topology by means of its graph $G(f) = \{(x, f(x)) : x \in X\}$. For convenience, we choose the following metric σ on $X \times Y$ which is compatible with the product topology.

$$\sigma[(x_1,y_1),(x_2,y_2)] = \max\{d(x_1,x_2), \rho(y_1,y_2)\}, \tag{4.1}$$

for all $x_1, x_2 \in X$ and for all $y_1, y_2 \in Y$. Thus in the above sense $C(X,Y) \subseteq CL(X \times Y)$, and hence $C(X,Y)$ can naturally inherit any hyperspace topology from $CL(X \times Y)$. For example, the Hausdorff metric on $C(X,Y)$ can be defined as follows:

$$H_\sigma(f,g) = \inf\{\varepsilon > 0 : G(g) \subseteq B(G(f),\varepsilon) \text{ and } G(f) \subseteq B(G(g),\varepsilon)\}$$

For details on Hausdorff metric topology on function spaces one may refer to [Holá (1988, 1992); Naimpally (1966)]. Let us note the following relation.

Proposition 4.1. *[Beer (1985)] Let (X,d) and (Y,ρ) be metric spaces. Then the Hausdorff metric topology on $C(X,Y)$ is weaker than the topology of uniform convergence on $C(X,Y)$.*

Proof. It can be easily verified that $H_\sigma(f,g) \leq d_{uc}(f,g)$ for all $f, g \in C(X,Y)$ and hence, the result follows. □

Let us denote the proximal topology and the Hausdorff metric topology on $C(X,Y)$ by τ_{δ_σ} and τ_{H_σ} respectively, where σ is the metric on $X \times Y$ defined by (4.1). We now conclude the following.

$$\tau_{\delta_\sigma} \subseteq \tau_{H_\sigma} \subseteq \tau_{uc}$$

In [Naimpally (1966)], it was proved that if we restrict the topologies τ_{δ_σ} and τ_{uc} to the set of uniformly continuous functions from X to Y, then the two topologies would be same.

Proposition 4.2. *[Beer (1993); Naimpally (1966)] Let (X,d) and (Y,ρ) be metric spaces. Then on the set of uniformly continuous functions from X to Y, $\tau_{\delta_\sigma} = \tau_{uc}$.*

Actually, the author in [Naimpally (1966)] proved the previous result in the following form: if a metric space (X,d) is compact and (Y,ρ) is any metric space, then $\tau_{H_\sigma} = \tau_{uc}$ on $C(X,Y)$. This equivalence plays a significant role in approximation theory, for instance see [Sendov (1990)]. In [Beer (1985)], it was proved that a metric space (X,d) is UC if and only if for any metric space (Y,ρ), $\tau_{H_\sigma} = \tau_{uc}$ on $C(X,Y)$. In the next section, we study the same equivalence on some particular subsets of the class of continuous functions and locally Lipschitz functions from (X,d) to some other metric space (Y,ρ). Interestingly, this also characterizes the cofinal completeness of the space (X,d). By trivial modification of some arguments in the proof of Proposition 4.2, one can see that on $C(X,Y)$, the *topology of pointwise convergence* is weaker than the proximal topology. We find it worthy to mention Lemma 2.3.2 of [Beer (1993)], which says that a net (f_λ) converges to f in $C(X,Y)$ with proximal topology if and only if $\lim_{\lambda} e_\sigma(G(f_\lambda), G(f)) = 0$, where $e_\sigma(A,B) = \inf\{\varepsilon > 0 : A \subseteq B(B,\varepsilon)\}$.

4.2 Cofinal Completeness vis-*à*-vis Hyperspaces

The purpose of this section is to give necessary and sufficient conditions for a metric space to be cofinally complete in terms of relations among some hyperspace and function space topologies. Along with a set of many significant internal characterizations of UC spaces in the literature, one can find characterizations of UC spaces in terms of hyperspace topologies and function space topologies as well [Beer *et al.* (1987); Beer (1985, 1993); Brandi *et al.* (2008)]. Let us recall the following well-known result.

Theorem 4.1. *[Atsuji (1958); Beer (1985)] For a metric space (X,d), the following assertions are equivalent.*

- (a) *(X,d) is UC.*
- (b) *On $CL(X)$, the Vietoris topology is weaker than the Hausdorff metric topology.*
- (c) *For every A and B in $CL(X)$, such that $A \cap B = \emptyset$, we have $D(A,B) > 0$.*

The equivalence of (a) and (b) and that of (a) and (c) can be found in [Beer (1985)] and [Atsuji (1958)] respectively. Since every UC space is cofinally complete, in this section we study a subset of $CL(X)$ such that the relation between the two hyperspace topologies characterizes cofinally complete metric spaces. For this purpose, we take the set of almost nowhere locally compact subsets of (X,d) denoted by $AC(X)$. Let us first recall the following result from Chapter 2, which acts as a fundamental property for the results presented in this section.

Theorem 4.2. *([Beer and Di Maio (2012)], Proposition 5.5) A metric space (X,d) is cofinally complete if and only if for every $A \in AC(X)$ and $B \in CL(X)$ with $A \cap B = \emptyset$, we have $D(A,B) > 0$.*

Though it seems natural that the relation given in Theorem 4.1 between the two hyperspace topologies will hold on $AC(X)$ for a cofinally complete metric space (X,d), we see that it is in fact sufficient for (X,d) to be cofinally complete. Throughout the section, we consider hyperspace topologies on the set $AC(X)$ which it inherits from the set $CL(X)$ as a subspace.

Remark 4.1. Observe that the space $AC(X)$ is not invariant under changes to an equivalent metric ρ on X. For example, consider the set of natural numbers. For all $n,m \in \mathbb{N}$, define $d_1(n,m) = |n-m|$ and $d_2(n,m) = |\frac{1}{n} - \frac{1}{m}|$. It can be easily verified that the set $AC(X)$ with respect to d_1 is precisely the collection of all non-empty finite subsets of $\mathbb{N}$, while the set $AC(X)$ with respect to d_2 is the collection of all non-empty subsets of $\mathbb{N}$.

Before moving forward, we would like to note a small observation which in some sense shows the challenge of dealing with $AC(X)$ in comparison to $CL(X)$.

Remark 4.2. Note that unlike closed sets, an almost nowhere locally compact subset need not be stable under small enlargements, that is, for a metric space (X,d), if $A \in AC(X)$, then there need not be any $\delta > 0$ such that the closure of $B(A,\delta)$ belongs to $AC(X)$. For example, consider $X = \{0\} \cup \{\frac{1}{j}e_n : j,n \in \mathbb{N}\}$ with the metric induced from the real Hilbert space l_2. Observe that $nlc(X) = \{0\}$. Clearly, the set $A = \{0\}$ is almost nowhere locally compact but there does not exist any $\delta > 0$ such that the closure of $B(A,\delta)$ is so.

Note that in the previous remark the space X is non-locally compact cofinally complete metric space. The following result shows that for a metric space (X,d), the condition: the Vietoris topology on $AC(X)$ is weaker than the Hausdorff metric topology on $AC(X)$, is not only necessary but also sufficient for (X,d) to be cofinally complete.

Theorem 4.3. *[Gupta and Kundu (2021a)] For a metric space (X,d), the following assertions are equivalent.*

(a) *(X,d) is cofinally complete.*
(b) *On $AC(X)$, $\tau_V \subseteq \tau_{H_d}$.*
(c) *On $AC(X)$, $\tau_{V^+} \subseteq \tau_{H_d}$.*

Proof. $(a) \Rightarrow (b)$: The subbasic open sets for the Vietoris topology on $AC(X)$ consist of the sets of the form: $W^+ = \{B \in AC(X) : B \subseteq W\}$, where W is an open set in X and $\mathscr{A}^- = \{B \in AC(X) : \text{for all } A \in \mathscr{A},\ B \cap A \neq \emptyset\}$, where $\mathscr{A}$ is a finite family of open sets in X. Let $A \in W^+$. Since W^c is closed, by Theorem 4.2, there exists $\varepsilon > 0$ such that $B(A,\varepsilon) \cap B(W^c,\varepsilon) = \emptyset$. It can be verified that $B_{H_d}(A,\varepsilon) \cap AC(X) \subseteq W^+$ and hence it follows that $W^+ \in \tau_{H_d}$.

Now consider $E \in \mathscr{A}^-$, where $\mathscr{A}$ consists of the open sets $A_1, A_2, \ldots, A_n$. Since $E \cap A_i \neq \emptyset$ for each $1 \leq i \leq n$, let $x_i \in E \cap A_i$ for each $1 \leq i \leq n$. Now, there exists $\varepsilon > 0$ such that $d(x_i, A_i^c) > \varepsilon$ for all i. Now let $D \in B_{H_d}(E,\varepsilon) \cap AC(X)$. Therefore, $E \subseteq B(D,\varepsilon)$ and thus for each x_i, there exists $y_i \in D$ such that $d(x_i,y_i) < \varepsilon$ for all i. Therefore for each i, $y_i \in D \cap A_i$ and hence $D \in \mathscr{A}^-$. We conclude that on $AC(X)$, $\tau_V \subseteq \tau_{H_d}$.

$(b) \Rightarrow (c)$: This implication is trivial.

$(c) \Rightarrow (a)$: Let $\tau_{V^+} \subseteq \tau_{H_d}$ on $AC(X)$. Suppose (X,d) is not cofinally complete. By Theorem 4.2, there exist $A \in AC(X)$ and $B \in CL(X)$ such that $A \cap B = \emptyset$, but $D(A,B) = 0$. Let $B^c = V$, thus $A \in V^+$. Since on $AC(X)$ $\tau_{V^+} \subseteq \tau_{H_d}$, there exists $n_o \in \mathbb{N}$ such that $B_{H_d}(A, \frac{1}{n_o}) \cap AC(X) \subseteq V^+$. Since $D(A,B) = 0$, for all $n \in \mathbb{N}$ there exist $a_n \in A$ and $b_n \in B$ such that $d(a_n,b_n) < \frac{1}{n}$. Since A and B are closed sets and $A \cap B = \emptyset$, the sequences (a_n) and (b_n) have no cluster points. Now for each $\varepsilon > 0$, the set $A_\varepsilon = \{a \in A\ :\ \nu(a) \geqslant \varepsilon\}$ is compact. Thus for all $\varepsilon > 0$, at most finitely many elements of the sequence (a_n) can belong to A_ε. Therefore, there exists a subsequence (a_{n_k}) of (a_n) such that $\lim_{n\to\infty} \nu(a_{n_k}) = 0$. Rename the sequence (a_{n_k}) to (a_n). By the uniform continuity of the functional ν, $\lim_{n\to\infty} \nu(b_n) = 0$. By passing to subsequences, we can assume that $\nu(a_n) < \frac{1}{n}$ and $\nu(b_n) < \frac{1}{n}$ for all $n \in \mathbb{N}$. Let $C = \{b_n : n > n_o\}$. Then $C \in AC(X)$ as for each $\varepsilon' > 0$, the set $\{x \in C\ :\ \nu(x) \geqslant \varepsilon'\}$ is finite. Therefore the set $A \cup C \in AC(X)$. Also, $A \cup C \subseteq B(A, \frac{1}{n_o})$ as for $b_k \in C$ $(k > n_o)$, there exists $a_k \in A$ such that $d(a_k,b_k) < \frac{1}{k} < \frac{1}{n_o}$. Therefore, $A \cup C \in B_{H_d}(A, \frac{1}{n_o}) \cap AC(X) \subseteq V^+ = (B^c)^+$. So $A \cup C \in (B^c)^+$, which is a contradiction. Hence (X,d) is cofinally complete. □

Remark 4.3. The implication $(a) \Rightarrow (b)$ in the previous theorem also follows from the fact that for a cofinally complete metric space (X,d), $A \in AC(X)$ implies that A is compact and thus $\tau_V \subseteq \tau_{H_d}$ on $AC(X)$ [Michael (1951), Theorem 3.3].

We have already mentioned that for a metric space (X,d), the proximal topology is weaker than the Vietoris topology on $CL(X)$. The next two results establish the condition under which the proximal topology is finer than the upper Vietoris topology and the proximal topology coincides with the Vietoris topology on $AC(X)$.

Theorem 4.4. *[Gupta and Kundu (2021a)] A metric space (X,d) is cofinally complete if and only if $\tau_{V^+} \subseteq \tau_{\delta_d}$ on $AC(X)$.*

Proof. Let (X,d) be a cofinally complete. Let W be an open set in X and let $A \in W^+$. Since (X,d) is cofinally complete, $D(A,W^c) > 0$. Thus for some $\delta > 0$, $A \subseteq B(A,\delta) \subseteq W$. It follows that $A \in (B(A,\delta))^{++} \cap AC(X) \subseteq W^+$. Hence on $AC(X)$, $W^+ \in \tau_{\delta_d}$.

Conversely, let the upper Vietoris topology on $AC(X)$ be weaker than the proximal topology on $AC(X)$. We know that on $CL(X)$, $\tau_{\delta_d} \subseteq \tau_{H_d}$. Since $AC(X) \subseteq CL(X)$, $\tau_{\delta_d} \subseteq \tau_{H_d}$ on $AC(X)$. Thus we get $\tau_{V^+} \subseteq \tau_{\delta_d} \subseteq \tau_{H_d}$ on $AC(X)$. Hence by Theorem 4.3, we get (X,d) is cofinally complete. □

Theorem 4.5. *[Gupta and Kundu (2021a)] A metric space (X,d) is cofinally complete if and only if $\tau_V = \tau_{\delta_d}$ on $AC(X)$.*

Proof. Let (X,d) be cofinally complete. Since $\tau_{\delta_d} \subseteq \tau_V$ on $CL(X)$, $\tau_{\delta_d} \subseteq \tau_V$ on $AC(X)$. Now to show $\tau_V \subseteq \tau_{\delta_d}$ on $AC(X)$, let $A \in W^+$ for some open set W in X. Since (X,d) is cofinally complete, $A \subseteq W$ implies $A \in W^{++}$. Thus $W^+ = W^{++} \cap AC(X)$ and hence $\tau_V \subseteq \tau_{\delta_d}$ on $AC(X)$.

Conversely, let $\tau_V = \tau_{\delta_d}$ on $AC(X)$. Suppose (X,d) is not cofinally complete. Using the same technique as in $(c) \Rightarrow (a)$ of Theorem 4.3, we can choose $A \in AC(X)$ and $B \in CL(X)$ such that $D(A,B) = 0$ and $A \cap B = \emptyset$. Also, there exist sequences $(a_n) \subseteq A$ and $(b_n) \subseteq B$ such that $d(a_n,b_n) < \frac{1}{n}$, $\nu(a_n) < \frac{1}{n}$ and $\nu(b_n) < \frac{1}{n}$ for all $n \in \mathbb{N}$. For every n, define $A_n = A \cup \{b_k : k \geq n\}$. Clearly, $A_n \in AC(X)$ for all $n \in \mathbb{N}$. Now we prove that the sequence (A_n) converges to A in proximal topology on $AC(X)$. Let $A \in W^{++}$ for some open subset W in X. Thus $A \subseteq B(A, \frac{1}{n_o}) \subseteq W$ for some $n_o \in \mathbb{N}$. We claim that $B(A_n, \frac{1}{3n}) \subseteq B(A, \frac{1}{n_o})$ for all $n \geqslant 2n_o$. For any $e \in B(A_n, \frac{1}{3n})$, if there exists $a \in A$ such that $d(e,a) < \frac{1}{3n}$, then $e \in B(A, \frac{1}{n_o})$. Otherwise there exists $b_t \in \{b_k : k \geq n\}$ such that $d(b_t, e) < \frac{1}{3n}$. Corresponding to this b_t, there exists $a_t \in A$ such that $d(b_t, a_t) < \frac{1}{t}$. Since $t \geq n$, $d(e,a_t) < \frac{1}{4n} < \frac{1}{n_o}$. Thus $B(A_n, \frac{1}{2n}) \subseteq B(A, \frac{1}{n_o}) \subseteq W$ for all $n \geqslant n_o$. Therefore, $A_n \in W^{++}$ for all $n \geqslant n_o$. Hence (A_n) converges to A in proximal topology on $AC(X)$. Now, $A \in (B^c)^+$, but $A_n \notin (B^c)^+$ for all $n \in \mathbb{N}$. Thus it follows that in the Vietoris topology, the sequence (A_n) does not converge to A. We get a contradiction. □

Let us see an example of a metric space which is not cofinally complete and thus the aforesaid relations among hyperspace topologies are not true.

Example 4.1. Consider the space $X = \bigcup_{n \in \mathbb{N}} A_n$, where $A_n = \{e_n + \frac{1}{n} e_k : k \in \mathbb{N}\}$ and $\{e_n : n \in \mathbb{N}\}$ is the standard orthonormal basis of l_2, as a metric subspace of the real Hilbert space l_2. Consider the sets $A = \{e_n + \frac{1}{n} e_1 : n \in \mathbb{N}\}$ and $B = \{e_n + \frac{1}{n} e_2 : n \in \mathbb{N}\}$. Clearly, both the sets A and B belong to $AC(X)$ as for all

$\varepsilon > 0$, the set $\{c \in A \cup B \ : \ \nu(c) \geqslant \varepsilon\}$ is finite. Since B is closed in X, the set $(B^c)^+ \in \tau_V$ on $AC(X)$. Let us now see that $(B^c)^+ \notin \tau_{\delta_d}$. For every $n \in \mathbb{N}$, let $A_n = A \cup \{e_t + \frac{1}{t}e_2 : t \geqslant n\}$. Clearly, $A_n \in AC(X)$ and (A_n) converges to A in proximal topology. Thus we have a sequence $(A_n) \subseteq ((B^c)^+)^c$ which does not converge in $((B^c)^+)^c$ as $A \in (B^c)^+$. Consequently, $((B^c)^+)^c$ is not closed and thus $(B^c)^+ \notin \tau_{\delta_d}$ on $AC(X)$. Hence on $AC(X)$, the proximal topology is strictly weaker than the Vietoris topology. By Theorem 4.4, we would like to note that the upper Vietoris topology is not weaker than the proximal topology on $AC(X)$.

In view of Theorem 4.3, let us observe that the set $(A^c)^+$ is open in Vietoris topology but $(A^c)^+$ is not open in Hausdorff topology as $B \in (A^c)^+$ but there does not exist any $\varepsilon > 0$ for which $B_{H_d}(B,\varepsilon) \subseteq (A^c)^+$. Thus the Hausdorff topology on $AC(X)$ is not finer than the Vietoris topology on $AC(X)$.

We have already mentioned that two metrics on a set X generate same Hausdorff metric topology on $CL(X)$ if and only if they are uniformly equivalent. Thus for a topological space X, we can endow $CL(X)$ with the supremum of all the Hausdorff metric topologies on $CL(X)$. Let us denote it by τ_{sup} on $CL(X)$, where

$$\tau_{sup} = \sup\{\tau_{H_d} \ : \ d \text{ metrizes } X\}$$

Let us recall the following interesting fact.

Proposition 4.3. *([Beer* et al. *(1987)], Theorem 2.1) Let X be a metrizable space. Then the locally finite topology on $CL(X)$ coincides with the supremum of all the Hausdorff metric topologies corresponding to equivalent metrics on X. That is, $\tau_{lf} = \tau_{sup}$.*

Lemma 4.1. *Let (X,d) be a metric space. Then,*

$$\tau_{sup}|_{AC(X)} = sup\{\tau_{H_{d'}}|_{AC(X)} \ : \ d' \ metrizes \ X\}$$

Proof. Let $V \in \tau_{H_{d'}}|_{AC(X)}$ for some metric d' on X equivalent to the metric d. Thus $V = W \cap AC(X)$ where W is an open set in $(CL(X), \tau_{H_{d'}})$. Consequently, W is an open subset of $(CL(X), \tau_{sup})$. Hence $V \in \tau_{sup}|_{AC(X)}$.

Conversely, let $V \in \tau_{sup}|_{AC(X)}$. Then $V = U \cap AC(X)$, where $U \in \tau_{sup}$ on $CL(X)$. We have $U = \bigcup_{i\in I}\bigcap_{j=1}^{n_i} V_{i_j}$, where each V_{i_j} is an open set in $(CL(X), H_d)$, where d metrizes X. Thus $V = \bigcup_{i\in I}\bigcap_{j=1}^{n_i}(V_{i_j} \cap AC(X))$ and hence $V \in sup\{\tau_{H_{d'}}|_{AC(X)} \ : \ d' \text{ metrizes } X\}$. □

Theorem 4.6. *[Gupta and Kundu (2021a)] A metric space (X,d) is cofinally complete if and only if $\tau_{H_d} = \tau_{lf}$ on $AC(X)$.*

Proof. Let the Hausdorff topology be same as the locally finite topology on $AC(X)$. If (X,d) is not cofinally complete, then by Theorem 4.3, there exists a subset of $AC(X)$ say U such that $U \in \tau_V$ but $U \notin \tau_{H_d}$, that is, $U \notin \tau_{lf}$. Since $\tau_V \subseteq \tau_{lf}$ on $AC(X)$, $U \notin \tau_V$. We get a contradiction.

Conversely, let (X,d) be cofinally complete. Since $AC(X) \subseteq CL(X)$, by Proposition 4.3, $\tau_{lf} = \tau_{sup}$ on $AC(X)$. Now let ρ be a metric on X which is equivalent to d. We claim that $\tau_{H_\rho} \subseteq \tau_{H_d}$ on $AC(X)$. Consider the following identity map.

$$I: \ (AC(X), H_d) \longrightarrow (AC(X), H_\rho)$$

To prove the claim, we need to prove that I is continuous. Let $F \in AC(X)$ and let $\varepsilon > 0$. Since ρ and d are equivalent metrics on X, the following identity function is continuous.

$$I': (X,d) \longrightarrow (X,\rho)$$

Since (X,d) is cofinally complete and $F \in AC(X)$, F is compact. Suppose for each $n \in \mathbb{N}$, there exist $x_n \in F$ and $y_n \in X$ such that $d(x_n, y_n) < \frac{1}{n}$ but $\rho(x_n, y_n) > \varepsilon$. Since F is compact, the sequence (x_n) has a cluster point say z. Thus z is a cluster point of the sequence (y_n) as well. This contradicts the continuity of I' at the point z. Thus for the given ε, there exists $\delta > 0$ such that if $x \in X$ and $y \in F$ with $d(x,y) < \delta$, then $\rho(x,y) < \varepsilon$. Now let $H \in B_{H_d}(F,\delta) \cap AC(X)$. Thus for $x \in F$, there exists $y \in H$ for which $d(x,y) < \delta$ and hence $\rho(x,y) < \varepsilon$. Consequently, $F \subseteq B_\rho(H,\varepsilon)$ and similarly, we can get that $H \subseteq B_\rho(F,\varepsilon)$. Thus $H_d(F,H) < \delta$ implies $H_\rho(F,H) < \varepsilon$, which proves the continuity of the map I. This shows that $\tau_{H_\rho} \subseteq \tau_{H_d}$ on $AC(X)$, and hence by Lemma 4.1, $\tau_{sup} = \tau_{H_d}$ on $AC(X)$. Since $\tau_{lf} = \tau_{sup}$ on $AC(X)$, $\tau_{lf} = \tau_{H_d}$ on $AC(X)$. □

Let us see an example of a metric space (X,d) in which $\tau_{H_d} \neq \tau_{lf}$ on $AC(X)$.

Example 4.2. Consider the metric space (X,d) defined in Example 4.1, where $A = \{e_n + \frac{1}{n}e_1 : n \in \mathbb{N}\}$. For each n, let $A_n = \{e_n + \frac{1}{n}e_1\}$. Clearly, $\mathscr{A} = \{A_n : n \in \mathbb{N}\}$ is a locally finite family of open sets in X and thus $\mathscr{A}^- \in \tau_{lf}$ on $AC(X)$. We now claim that $\mathscr{A}^-$ is not open in τ_{H_d} on $AC(X)$. Let $B_n = A \setminus A_n \cup \{e_n + \frac{1}{n}e_2\}$. Clearly, $B_n \in AC(X) \cap (\mathscr{A}^-)^c$. Let us observe that for each $\varepsilon > 0$, there exists $n_o \in \mathbb{N}$ such that $H_d(B_n, A) < \varepsilon$ for all $n \geqslant n_o$. Thus (B_n) converges to A in Hausdorff metric topology, but $A \in \mathscr{A}^-$, which implies that $(\mathscr{A}^-)^c$ is not closed in the Hausdorff metric on the subspace $AC(X)$.

Note that for a UC space (X,d), it is not necessary that $CL(X) = AC(X)$ or vice-versa. For example, consider $X = \{\frac{1}{n} : n \in \mathbb{N}\}$ with the usual distance metric, then (X,d) is not UC but $AC(X) = CL(X)$. Also, considering the metric space of Remark 4.2, one can see that (X,d) is a UC space but $AC(X) \neq CL(X)$ because choosing $\delta > 0$ small enough, we get $\overline{B(0,\delta)}$ to be closed but $\overline{B(0,\delta)} \notin AC(X)$. The following example shows that for a cofinally complete metric space (X,d), the Vietoris topology and the proximal topology need not be same on $CL(X)$.

Example 4.3. Let $A = \{n : n \in \mathbb{N}\}$, clearly $A \in CL(\mathbb{R})$ and $(A^c)^+$ is open in Vietoris topology. Since $B_n = \{n + \frac{1}{n} : n \in \mathbb{N}\} \cup \{k : k \geq n,\ k \in \mathbb{N}\}$ belongs to $((A^c)^+)^c$ for each n and the sequence (B_n) converges to $B = \{n + \frac{1}{n} : n \in \mathbb{N}\} \in (A^c)^+$ with respect to the proximal topology. Hence $(A^c)^+$ is not open in τ_{δ_d} on $CL(X)$.

It is important to note that in the previous example, for each $n \in \mathbb{N}$, $B_n \in CL(\mathbb{R})$ but $B_n \notin AC(\mathbb{R})$. Next, we study some necessary and sufficient conditions for a metrizable topological space X to admit a cofinally complete metric. Observe that since ν is a continuous function for a metric space (X,d), $nlc(X)$ is always closed in X. And hence in the following result, we consider hyperspace topologies on the set $CL(nlc(X))$ which it inherits from the space $CL(X)$ as a subspace.

Theorem 4.7. *[Gupta and Kundu (2021a)] For a metrizable topological space X, the following statements are equivalent.*

- (a) *A compatible metric d on X exists such that the subset $nlc(X)$ of X is compact.*
- (b) *A compatible metric d on X exists such that (X,d) is cofinally complete.*
- (c) *A compatible metric d on X exists for which τ_{lf} on the subspace $AC(X)$ is metrizable.*
- (d) *A compatible metric d on X exists for which τ_{lf} on the subspace $AC(X)$ is first countable.*
- (e) *A compatible metric d on X exists for which τ_{V^+} on the subspace $AC(X)$ is first countable.*
- (f) *A compatible metric d on X exists for which τ_V on the subspace $CL(nlc(X))$ is second countable.*
- (g) *A compatible metric d on X exists for which τ_V on the subspace $CL(nlc(X))$ is first countable.*
- (h) *A compatible metric d on X exists for which τ_{V^+} on the subspace $CL(nlc(X))$ is second countable.*
- (i) *A compatible metric d on X exists for which τ_{V^+} on the subspace $CL(nlc(X))$ is first countable.*

Proof. The equivalence of (a) and (b) follows from Theorem 2.6. The implications $(c) \Rightarrow (d)$, $(f) \Rightarrow (g)$ and $(h) \Rightarrow (i)$ are easy to follow.

$(b) \Rightarrow (c)$: Let d be a compatible metric on X which makes (X,d) a cofinally complete space. By Theorem 4.6, τ_{lf} and τ_{H_d} are same on $AC(X)$. Thus τ_{lf} on the subspace $AC(X)$ is metrizable.

$(d) \Rightarrow (a)$: Suppose $(nlc(X), d_1)$ is not compact for every compatible metric d_1 on X. Let d be a compatible metric on X given by (d). Then there exists a sequence of distinct points, $(x_n) \subseteq nlc(X)$ such that it does not cluster in X. Let $A = \{x_n \ : \ n \in \mathbb{N}\}$. Clearly, $A \in AC(X)$. Let $\mathscr{W}_n = W_n^+ \cap \mathscr{A}_n^-$ be a countable base for $A \in (AC(X), \tau_{lf})$, where each W_n is an open set in X and each $\mathscr{A}_n$ is a locally finite family of open sets in X. Let $\{B(x_n, \varepsilon_n) : n \in \mathbb{N}\}$ be a family of pairwise disjoint open sets. Now, $A \subseteq W_n$ for all $n \in \mathbb{N}$ and therefore, $x_n \in W_n$ for all $n \in \mathbb{N}$. Since $\nu(x_n) = 0$, there exists $y_n \in W_n$ such that $0 < d(x_n, y_n) < \frac{\varepsilon_n}{2}$ for all $n \in \mathbb{N}$. Let $O = \bigcup_{n \in \mathbb{N}} B(x_n, \varepsilon_n) \setminus \{y_n\}$. Clearly, $A \in O^+$. Therefore, there exists $n_o \in \mathbb{N}$ such that $\mathscr{W}_{n_o} \subseteq O^+$. We get a contradiction because $A \cup \{y_{n_o}\} \in AC(X)$ which belongs to $\mathscr{W}_{n_o}$, but $A \cup \{y_{n_o}\} \notin O^+$. Thus it follows that $(nlc(X), d)$ is compact.

$(b) \Rightarrow (e)$: Let (X,d) be cofinally complete for some compatible metric d on X. For each $A \in AC(X)$, if $A \subseteq O$ for some open set O in X, then $D(A, O^c) > 0$, that is, $B(A, \frac{1}{n}) \subseteq O$ for some $n \in \mathbb{N}$. Thus $\{(B(A, \frac{1}{n}))^+ \ : \ n \in \mathbb{N}\}$ is a countable base at $A \in AC(X)$ with respect to τ_{V^+} on $AC(X)$.

$(a) \Rightarrow (f)$: Since $nlc(X)$ is a compact subset of (X,d), there exists a countable base of topology on $nlc(X)$ inherited from X, say $\{U_1, U_2, U_3, \ldots\}$, where each $U_i = V_i \cap nlc(X)$ for some open subset V_i of X. We know that the subbasic open sets for the Vietoris topology on $CL(nlc(X))$ are of the form $O^+ \cap CL(nlc(X))$ (O is open in X) and $\mathscr{A}^- \cap CL(nlc(X))$ ($\mathscr{A}$ is a finite family of open sets in X). It can be verified that the following countable subbase generates the same topology on $CL(nlc(X))$.

$$\mathscr{B} = \left\{ \Big(\bigcup_{i \in \mathscr{F}} U_i \Big)^+, \mathscr{A}_{\mathscr{F}}^- : \mathscr{F} \text{ is a finite subset of } \mathbb{N} \right\}$$

where $\mathscr{A}_{\mathscr{F}} = \{U_i : i \in \mathscr{F}\}$. Thus the inherited Vietoris topology on $CL(nlc(X))$ is second countable.

The implication $(a) \Rightarrow (h)$ follows from the proof of $(a) \Rightarrow (f)$. The implications $(g) \Rightarrow (a)$, $(i) \Rightarrow (a)$ and $(e) \Rightarrow (a)$ can be proved in a manner similar to the proof of the implication $(d) \Rightarrow (a)$. □

For metric spaces (X,d) and (Y,ρ), we next aim to find a subspace of $C(X,Y)$ on which the proximal topology is same as the topology of uniform convergence so that this equivalence characterizes the cofinal completeness of (X,d). For this purpose, we consider the subset $CV(X,Y)$ of $C(X,Y)$, which has already been considered in Chapter 2. Analogous to this subset of continuous function, we consider a similar subset of locally Lipschitz functions as well. Let us denote the set of locally Lipschitz functions between two metrics spaces (X,d) and (Y,ρ) by $LL(X,Y)$ and let the set $LLV(X,Y)$ be $LL(X,Y) \cap CV(X,Y)$. Now we consider hyperspace topologies on the sets $CV(X,Y)$ and $LLV(X,Y)$, which they inherit from the space $C(X,Y)$ as subspaces. The set $CV(X,\mathbb{R})$ ($LLV(X,\mathbb{R})$) is denoted by $CV(X)$ ($LLV(X)$). Further, the set of all bounded functions in $CV(X)$ is denoted by $CV^*(X)$.

Theorem 4.8. *[Gupta and Kundu (2021a)] For a metric space (X,d), the following assertions are equivalent.*

(a) *(X,d) is cofinally complete.*
(b) *Whenever (Y,ρ) is a metric space, then $\tau_{\delta_\sigma} = \tau_{uc}$ on $CV(X,Y)$.*
(c) *On $CV(X)$, $\tau_{\delta_\sigma} = \tau_{uc}$.*
(d) *On $CV^*(X)$, $\tau_{\delta_\sigma} = \tau_{uc}$.*
(e) *There exists a metric space (Y,ρ) containing a nonconstant path such that $\tau_{uc} = \tau_{\delta_\sigma}$ on $CV(X,Y)$.*
(f) *There exists a metric space (Y,ρ) containing a nonconstant path such that $\tau_{H_\sigma} = \tau_{uc}$ on $CV(X,Y)$.*
(g) *Whenever, (Y,ρ) is a metric space, then $\tau_{H_\sigma} = \tau_{uc}$ on $CV(X,Y)$.*
(h) *On $CV(X)$, $\tau_{H_\sigma} = \tau_{uc}$.*
(i) *On $CV^*(X)$, $\tau_{H_\sigma} = \tau_{uc}$.*
(j) *Whenever (Y,ρ) is a metric space, then $\tau_{\delta_\sigma} = \tau_{uc}$ on $LLV(X,Y)$.*
(k) *On $LLV(X)$, $\tau_{\delta_\sigma} = \tau_{uc}$.*
(l) *On $LLV^*(X)$, $\tau_{\delta_\sigma} = \tau_{uc}$.*
(m) *On $LLV^*(X)$, $\tau_{H_\sigma} = \tau_{uc}$.*

Proof. The implications $(b) \Rightarrow (c) \Rightarrow (d)$, $(g) \Rightarrow (h) \Rightarrow (i)$ and $(b) \Rightarrow (j) \Rightarrow (k) \Rightarrow (l)$ are all immediate. Since $CV(X,[0,1]) \subseteq CV^*(X)$, $(d) \Rightarrow (e)$ and $(i) \Rightarrow (f)$ are easy to see.

$(a) \Rightarrow (b)$: Since (X,d) is cofinally complete, by Theorem 2.10, $f \in CV(X)$ implies f is uniformly continuous. Now the result follows from Proposition 4.2.

$(e) \Rightarrow (f)$, $(b) \Rightarrow (g)$ and $(l) \Rightarrow (m)$: For a metric space (Y,ρ), $LV(X,Y) \subseteq CV(X,Y) \subseteq C(X,Y) \subseteq CL(X \times Y)$ and we already know that $\tau_{\delta_\sigma} \subseteq \tau_{H_\sigma} \subseteq \tau_{uc}$ on $C(X,Y)$. Thus the implications follow.

$(f) \Rightarrow (a)$: Suppose (X,d) is not cofinally complete. By Theorem 2.10, there exists $f \in CV(X,Y)$ such that f is not uniformly continuous on (X,d). Thus there exists $\lambda > 0$ and sequences (x_n) and (y_n) such that $d(x_n,y_n) < \frac{1}{n}$ but

$\rho(f(x_n), f(y_n)) > \lambda$ for all $n \in \mathbb{N}$. Since f is continuous, the sequences (x_n) and (y_n) do not cluster. Clearly, the case: $\inf_{n\in\mathbb{N}} \nu(x_n) > 0$ and $\inf_{n\in\mathbb{N}} \nu(y_n) > 0$ is not possible. Let $\inf_{n\in\mathbb{N}} \nu(x_n) = 0$. Since ν is uniformly continuous, $\inf_{n\in\mathbb{N}} \nu(y_n) = 0$. Thus there exists a subsequence (x_{n_k}) of the sequence (x_n) such that $\lim_{k\to\infty} \nu(x_{n_k}) = 0$ and hence $\lim_{k\to\infty} \nu(y_{n_k}) = 0$. Rename the sequences (x_{n_k}) and (y_{n_k}) to (x_n) and (y_n) respectively.

Case 1: For all n, x_n and y_n are isolated points in (X,d).
For each $n \in \mathbb{N}$, define $f_n : (X,d) \to (Y,\rho)$ by

$$f_n(z) = \begin{cases} f(y_n) & : \text{ if } z = x_n \\ f(x_n) & : \text{ if } z = y_n \\ f(z) & : \text{ otherwise} \end{cases}$$

Note that $f_n \in CV(X,Y)$ for all $n \in \mathbb{N}$. Also, for all n, $d_{uc}(f_n, f) > \lambda$ and $H_\sigma(f_n, f) \leq \frac{1}{n}$. This shows that $(f_n) \to f$ in $(CV(X,Y), H_\sigma)$, but (f_n) does not converge to f in $(CV(X,Y), \tau_{uc})$. We get a contradiction.

Case 2: One of the sequences (x_n) and (y_n) has infinitely many distinct terms that are limit points of X. Without loss of generality, assume that for each n, x_n is a limit point. Since the sequence (x_n) has no cluster point, for each $n \in \mathbb{N}$, there exists ε_n such that $0 < \varepsilon_n < \frac{1}{n}$ and the family of open balls $\{B(x_n, \varepsilon_n) : n \in \mathbb{N}\}$ is pairwise disjoint. Since x_n is a limit point, choose n distinct points $x_{1n}, x_{2n}, \ldots, x_{nn}$ in $B(x_n, \varepsilon_n)$ for each $n \in \mathbb{N}$. Let $\varepsilon_{1n}, \varepsilon_{2n}, \ldots, \varepsilon_{nn}$ be positive real numbers such that the family of open balls $\{B(x_{kn}, \varepsilon_{kn}) : 1 \leq k \leq n\}$ is pairwise disjoint and $\bigcup_{k=1}^{n} B(x_{kn}, \varepsilon_{kn}) \subseteq B(x_n, \varepsilon_n)$. Define $t_n : (X,d) \to [0,1]$ by

$$t_n(z) = \begin{cases} \frac{k}{n}\left(1 - \frac{d(z, x_{kn})}{\varepsilon_{kn}}\right) & : z \in B\left(x_{kn}, \varepsilon_{kn}\right) \text{ for some } 1 \leq k \leq n \\ 0 & : \text{ otherwise} \end{cases}$$

It is easy to see that t_n is Lipschitz continuous for each n. Define $t : (X,d) \to [0,1]$ by

$$t(z) = \begin{cases} t_n(z) & : z \in B(x_n, \varepsilon_n) \text{ for some } n \in \mathbb{N} \\ 0 & : \text{ otherwise} \end{cases}$$

It can be verified that $t \in CV(X)$. Since (Y,ρ) contains a nonconstant path, there exists a uniformly continuous function $\phi : [0,1] \to (Y,\rho)$ such that $\phi(0) \neq \phi(1)$. Let $\phi(0) = a$ and $\phi(1) = b$. Define the following continuous function.

$$g : (X,d) \longrightarrow (Y,\rho)$$

$$z \longmapsto \phi(t(z))$$

We claim that $g \in CV(X,Y)$. Let $\varepsilon > 0$. Since $\lim_{n\to\infty} \nu(x_n) = 0$ and ν is uniformly continuous, there exists $m \in \mathbb{N}$ such that $\{x \in X : \nu(x) > \varepsilon\} \cap \bigcup_{n=m}^{\infty} B(x_n, \varepsilon_n) = \emptyset$. Since t_n is uniformly continuous for each n, t restricted to $\left(\bigcup_{n=m}^{\infty} B(x_n,\varepsilon_n)\right)^c$ is uniformly continuous and g being a composition of two uniformly continuous functions, is also uniformly continuous on the set $\{x \in X : \nu(x) > \varepsilon\}$. Thus $g \in CV(X,Y)$.

Now for each $n \in \mathbb{N}$, define $h'_n : (X,d) \to [0,1]$ by

$$h'_n(z) = \begin{cases} (1-\frac{k}{n})\left(1-\frac{d(z,x_{kn})}{\varepsilon_{kn}}\right) & : z \in B\left(x_{kn},\varepsilon_{kn}\right) \text{ for some } 1 \leq k \leq n \\ 0 & : \text{otherwise} \end{cases}$$

and $h_n : (X,d) \to [0,1]$ by

$$h_n(z) = \begin{cases} h'_n(z) & : z \in B(x_n,\varepsilon_n) \\ t(z) & : \text{otherwise} \end{cases}$$

Define $g_n : (X,d) \to (Y,\rho)$ by $g_n(z) = \phi(h_n(z))$. By similar arguments (that were used for g) we can see that $g_n \in CV(X,Y)$.

Now, $g_n(x_{nn}) = \phi(0)$ and $g(x_{nn}) = \phi(1)$. Thus $d_{uc}(g_n,g) \geqslant \rho(a,b)$, which implies $(g_n) \nrightarrow (g)$ in $(CV(X,Y),d_{uc})$. Now we claim that $(g_n) \to (g)$ in $(CV(X,Y),H_\sigma)$.

Let $\varepsilon > 0$. We will show that there exists $n_o \in \mathbb{N}$ such that $H_\sigma(g,g_n) < \varepsilon\ \forall n \geq n_o$, that is, $G(g_n) \subseteq B(G(g),\varepsilon)\ \forall n \geq n_o$ and $G(g) \subseteq B(G(g_n),\varepsilon)\ \forall n \geq n_o$. We will prove the first inclusion, the proof of the second one is similar. Since ϕ is uniformly continuous, there exists $n_o \in \mathbb{N}$ such that $|x-y| < \frac{1}{n_o}$ implies $\rho(\phi(x),\phi(y)) < \varepsilon$. Choose n_o such that $n_o > \frac{2}{\varepsilon}$. Let $n \in \mathbb{N}$ be such that $n \geq n_o$. If $z \notin B(x_n,\varepsilon_n)$, then $(z,g_n(z))$ lies on the graph of g and we have nothing to prove. If $z \in B(x_n,\varepsilon_n)$, then there exists $p \in \mathbb{N}$, $1 \leq p \leq n$, such that $|h_n(z) - \frac{p}{n}| < \frac{1}{n} < \frac{1}{n_o}$. Thus we have $|h_n(z) - t(x_{pn})| < \frac{1}{n} < \frac{1}{n_o}$ which implies $\rho(\phi(h_n(z)),\phi(t(x_{pn})) < \varepsilon$. As a result, $\sigma[(z,g_n(z)),(x_{pn},g(x_{pn}))] = \max\{\frac{2}{n},\varepsilon\} = \varepsilon$. Hence $G(g_n) \subseteq B(G(g),\varepsilon)$ for all $n \geq n_o$ and we are done.

$(m) \Rightarrow (a)$: By observing that the functions h_n, t defined in the implication $(f) \Rightarrow (a)$ belong to $LLV^*(X)$, we can prove $(m) \Rightarrow (a)$ in a manner similar to the proof of $(f) \Rightarrow (a)$. □

4.3 Cofinal Completeness of the Space $(AC(X),H_d)$

In [Beer and Di Maio (2010)], for a metric space (X,d), Beer and Di Maio carried out an interesting study of cofinal completeness of the space $(CL(X),H_d)$.

They proved that the hyperspace $(CL(X), H_d)$ is cofinally complete if and only if X has a compact neighborhood in $(CL(X), H_d)$ (see Example 4.10). In the previous section, we have seen that $AC(X)$ plays a big role in the study of cofinal completeness of the space (X, d). Thus this section is devoted to study the cofinal completeness of the space $(AC(X), H_d)$. To achieve the intended goal, we first recall the following result from [Beer (1993)].

Proposition 4.4. *([Beer (1993)], Theorem 3.2.4) Let (X,d) be a metric space. Then*

(a) *$(CL(X), H_d)$ is complete if and only if (X,d) is complete.*
(b) *$(CL(X), H_d)$ is totally bounded if and only if (X,d) is totally bounded.*
(c) *$(CL(X), H_d)$ is compact if and only if (X,d) is compact.*

Theorem 4.9. *[Gupta and Kundu (2021a)] Let (X,d) be a metric space. Then*

(a) *$(AC(X), H_d)$ is complete if and only if (X,d) is complete.*
(b) *$(AC(X), H_d)$ is totally bounded if and only if (X,d) is totally bounded.*
(c) *$(AC(X), H_d)$ is compact if and only if (X,d) is compact.*

Proof. The necessity factor in (a), (b) and (c) follow from the fact that $\phi : (X,d) \to (CL(X), H_d)$ defined by $\phi(x) = \{x\}$ for all $x \in X$ is isometry and since the set $\{\{x\} : x \in X\}$ is closed in $(CL(X), H_d)$, it is closed in $(AC(X), H_d)$ as well.

(a): Let (X,d) be complete. By Proposition 4.4, $(CL(X), H_d)$ is also complete. Thus it is sufficient to prove that $(AC(X), H_d)$ is a closed subset of $(CL(X), H_d)$. Let (A_n) be a sequence in $AC(X)$ such that $(A_n) \to A \in CL(X)$ in τ_{H_d}. To prove that $A \in AC(X)$, let $\varepsilon > 0$. Consider the set $Z = \{a \in A : \nu(a) \geqslant \varepsilon\}$ and let $(a_n) \subseteq Z$. Suppose (a_n) does not have any cluster point, then since Z is complete, there exists $\delta > 0$ for which $\{B(a_n, \delta) : n \in \mathbb{N}\}$ is a disjoint family of open sets. Choose $0 < \lambda < \min\{\frac{\delta}{2}, \frac{\varepsilon}{2}\}$. Since $(A_n) \to A$, there exists $n_o \in \mathbb{N}$ such that $A \subseteq B(A_n, \lambda)$ for all $n \geq n_o$. Thus there exists a sequence $(y_n) \subseteq A_{n_o}$ such that $d(a_n, y_n) < \lambda$. Since $\nu(a_n) \geqslant \varepsilon$ and $d(a_n, y_n) < \frac{\varepsilon}{2}$, $\nu(y_n) \geqslant \frac{\varepsilon}{2}$ for all $n \in \mathbb{N}$. Also because $\lambda < \frac{\delta}{2}$, $\{B(y_n, \frac{\delta}{2}) : n \in \mathbb{N}\}$ is a disjoint family of open sets. We get contradiction to the fact that $A_{n_o} \in AC(X)$. Thus Z is compact, which implies that $A \in AC(X)$. Hence $(AC(X), H_d)$ is complete.

(b): Let (X,d) be totally bounded. By Proposition 4.4, $(CL(X), H_d)$ is totally bounded and hence so is $(AC(X), H_d)$.

(c): It follows from (a) and (b). □

The next theorem characterizes the cofinal completeness of the space $(AC(X), H_d)$ in terms of the uniform local compactness of (X,d).

Theorem 4.10. *[Gupta and Kundu (2021a)] For a metric space (X,d), the following assertions are equivalent.*

(a) *(X,d) is uniformly locally compact.*
(b) *$(AC(X),H_d)$ is uniformly locally compact.*
(c) *$(AC(X),H_d)$ is cofinally complete.*

Proof. $(a) \Rightarrow (b)$: Let $\alpha > 0$ such that $\nu(x) > \alpha$ for all $x \in X$ and let $B \in AC(X)$. Note that B is compact since it coincides with the compact set $\{x \in B : \nu(x) \geq \alpha\}$. Let $0 < \mu < \frac{\alpha}{2}$, then $\{A \in AC(X) : H_d(A,B) \leq \mu\}$ is a closed subset of $(AC(E),H_d)$, where $E = \{x \in X : d(x,B) \leq \mu\}$. By Theorem 4.9, we finish if we prove that E is compact. Indeed, since B is compact, there exists $\{x_1,x_2,\dots,x_n\} \subseteq B$ such that $B \subseteq \bigcup_{i=1}^{n} B(x_i,\frac{\alpha}{2})$. Thus $E \subseteq B(B,\frac{\alpha}{2}) \subseteq \bigcup_{i=1}^{n} B(x_i,\frac{\alpha}{2}) \subseteq \bigcup_{i=1}^{n} \overline{B(x_i,\alpha)}$. Since $\nu(x) > \alpha$ for all $x \in X$, E is compact.

$(b) \Rightarrow (c)$: This is immediate.

$(c) \Rightarrow (a)$: Let $(AC(X),H_d)$ be cofinally complete. Therefore, (X,d) is also cofinally complete. If $nlc(X)$ is empty, then X is uniformly locally compact because if (x_n) is a sequence in X such that $\lim_{n\to\infty} \nu(x_n) = 0$, then (x_n) would have a cluster point which further would lie in $nlc(X)$. If $nlc(X) \neq \emptyset$, then by Theorem 2.7, $nlc(X)$ is compact. Since (X,d) is a non-compact complete metric space, it is not totally bounded. Also the compactness of $nlc(X)$ implies that $(nlc(X))^c$ is not totally bounded. So there exist $\delta > 0$ and a sequence $(x_n) \subseteq (nlc(X))^c$ such that $\{B(x_n,\delta) : n \in \mathbb{N}\}$ is a disjoint family of open sets. Since $A = \{x_n : n \in \mathbb{N}\}$ is a closed set and $nlc(X)$ is compact, $D(A,nlc(X)) > \alpha > 0$. Let $z \in nlc(X)$ and $A_n = \{z,x_n\}$. We claim that $\nu(A_n) = 0$ in $(AC(X),H_d)$ for each $n \in \mathbb{N}$. Fix $n \in \mathbb{N}$ and let $\varepsilon > 0$. Our aim is to show that $\nu(A_n) < \varepsilon$. Since $\overline{B(z,\frac{\varepsilon}{2})}$ is non-compact, there exists a sequence $(z_t) \subseteq B(z,\varepsilon)$ with no cluster point. For each $t \in \mathbb{N}$, let $B_t = A_n \cup \{z_t\}$. It is clear that for each $t \in \mathbb{N}$, $B_t \in AC(X)$ and $H_d(A_n,B_t) < \varepsilon$. Now (B_t) does not have any cluster point because otherwise (z_t) would have a cluster point too. Thus $\nu(A_n) < \varepsilon$ for all $\varepsilon > 0$ and consequently, $\nu(A_n) = 0$ for all $n \in \mathbb{N}$.

Now $H_d(A_n,A_m) \geqslant \min\{\delta,\alpha\}$ for all $n \neq m$ and thus (A_n) has no cluster point. We get contradiction to the fact that $(AC(X),H_d)$ is cofinally complete. Hence (X,d) is uniformly locally compact. □

Remark 4.4. The implication $(c) \Rightarrow (a)$ of the previous theorem also follows from Corollary 3.5 in [Künzi and Romaguera (1999)] (see Exercise 4.11).

Before moving forward, let us make the following observation.

Proposition 4.5. *Let $A \in AT(X)$ in a metric space (X,d). Then $\widehat{A} \in AC(\widehat{X})$.*

Proof. Let $A \in AT(X)$. To show that $\widehat{A} \in AC(\widehat{X})$, let $\varepsilon > 0$ and let $Z = \{a \in \widehat{A} : \hat{v}(a) \geqslant \varepsilon\}$. Suppose Z is not compact. Since Z is a closed subset of a complete set, there exists a sequence $(x_n) \subseteq Z$ and $\delta > 0$ such that $\{B(x_n, 2\delta) : n \in \mathbb{N}\}$ is a disjoint family of open balls. Let $\lambda = \min\{\delta, \frac{\varepsilon}{2}\}$. Since A is dense in $\widehat{A}$, there exists a sequence (y_n) in A such that for each n, $y_n \in B(x_n, \lambda)$. Since $d(x_n, y_n) < \frac{\varepsilon}{2}$, $B(y_n, \frac{\varepsilon}{2}) \subseteq B(x_n, \varepsilon)$ and thus $\hat{v}(y_n) \geqslant \frac{\varepsilon}{2}$. Thus we have $t(y_n) = \hat{v}(y_n) \geqslant \frac{\varepsilon}{2}$. This is a contradiction to the fact that $A \in AT(X)$ because (y_n) is a sequence of A such that $\{B(y_n, \delta) : n \in \mathbb{N}\}$ is a disjoint family of open balls and $t(y_n) \geqslant \frac{\varepsilon}{2}$. □

In [Jain and Kundu (2008)], the authors proved that for a metric space (X,d), the completion of $(CL(X), H_d)$ is nothing but $(CL(\widehat{X}), H_d)$. Similarly, in view of Proposition 4.5, one may think that the completion of $(AT(X), H_d)$ is $(AC(\widehat{X}), H_d)$. But the following example shows that this need not be true in general. However, under some special conditions, the completion of $(AT(X), H_d)$ is $(AC(\widehat{X}), H_d)$.

Example 4.4. Consider $Y = \left\{\frac{1}{j}e_n : j, n \in \mathbb{N}\right\}$ with the metric induced from the real Hilbert space l_2. Let X be a space containing uncountably many copies of Y. More precisely, for each $\lambda \in I = [0,1]$, let (X, λ) be a copy of Y. Denote (X, λ) by X_λ and denote any element (x, λ) of (X, λ) by x_λ. Now $X = \bigcup_{\lambda \in I} X_\lambda$ is a metric space where the metric d on X is defined as follows.

$$d(x_\lambda, y_\alpha) = \begin{cases} |x - y| & : \lambda = \alpha \\ 2 & : \text{otherwise} \end{cases}$$

The completion $\widehat{X}$ of X is $X \cup \{a_\lambda : \lambda \in I\}$, where a_λ represents 0 of each copy X_λ. Moreover, $A = \{a_\lambda : \lambda \in I\} \in AC(\widehat{X})$ as each member of A is a point of non-local compactness. We claim that for any $0 < \varepsilon < 1$, there does not exist any $B \in AT(X)$ such that $\widehat{B} \subseteq H_d(A, \varepsilon)$. Suppose the claim is not true, that is, there exists $B \in AT(X)$ such that $\widehat{B} \subseteq H_d(A, \varepsilon)$. Since X is discrete, it is locally totally bounded and so is B. Now

$$\{x \in B : t(x) > 0\} = \bigcup_{n \in \mathbb{N}} \left\{x \in B : t(x) \geqslant \frac{1}{n}\right\}$$

If each set on the right hand side is countable, then B would be countable but that is not possible. Thus there exists $n_o \in \mathbb{N}$ such that for uncountably many elements of B, $t(x) \geqslant \frac{1}{n_o}$. This shows that there exists uncountably many elements of B each from a different copy with $t(x) \geqslant \frac{1}{n_o}$. Since these elements cannot have any Cauchy subsequence, we get a contradiction.

Theorem 4.11. *[Gupta and Kundu (2021a)] Let (X,d) be a metric space such that $(\widehat{X},d)$ is cofinally complete. Then the completion of $(AT(X),H_d)$ is $(AC(\widehat{X}),H_d)$.*

Proof. By Proposition 4.5, we can define the following function.

$$\phi : (AT(X),H_d) \longrightarrow (AC(\widehat{X}),H_d)$$

$$A \longmapsto \widehat{A}$$

Since $\sup_{a\in A} d(a,B) = \sup_{a\in \widehat{A}} d(a,\widehat{B})$ for all $A,B \subseteq X$, ϕ is an isometry. It remains to show that $\phi(AT(X))$ is dense in $AC(\widehat{X})$. Let $A \in AC(\widehat{X})$ and let $\varepsilon > 0$. Since $\widehat{X}$ is cofinally complete, A is compact. Thus for $\varepsilon > 0$, there exists $\bigcup_{i=1}^{n}\{x_i\} \subseteq A$ such that $A \subseteq \bigcup_{i=1}^{n} B(x_i,\frac{\varepsilon}{2})$. For each i, choose $y_i \in B(x_i,\frac{\varepsilon}{2})$ such that $y_i \in X$. Let $B = \{y_i : 1 \le i \le n\}$. Clearly, $B \in AT(X)$ and $\phi(B) = B \in B_{H_d}(A,\varepsilon)$. Hence $(AC(\widehat{X}),H_d)$ is the completion of $(AT(X),H_d)$. □

Now the natural question arises: When is the completion of $(AT(X),H_d)$ cofinally complete? The following result answers the same.

Theorem 4.12. *[Gupta and Kundu (2021a)] For a metric space (X,d), the following assertions are equivalent.*

(a) *(X,d) is uniformly locally totally bounded.*
(b) *The completion $(\widehat{X},d)$ of (X,d) is uniformly locally totally bounded.*
(c) *The completion $(\widehat{X},d)$ of (X,d) is uniformly locally compact.*
(d) *$(AC(\widehat{X}),H_d)$ is uniformly locally compact.*
(e) *$(AC(\widehat{X}),H_d)$ is uniformly locally totally bounded.*
(f) *$(AC(\widehat{X}),H_d)$ is cofinally complete.*
(g) *$(CAT(X),H_d)$ is uniformly locally totally bounded.*
(h) *The completion of $(AT(X),H_d)$ is cofinally complete.*

Proof. Since $(\widehat{X},d)$ is complete, the statements (b) and (c) are equivalent and similarly by Theorem 4.9, the statements (d) and (e) are equivalent. The equivalence of (c), (d) and (f) follows from Theorem 4.10.

$(a) \Rightarrow (b)$: Let $t(x) > \alpha > 0$ for all $x \in X$. Let $\hat{x} \in \widehat{X}$ and let (x_n) be a sequence in $\widehat{X}$ such that $(x_n) \subseteq B(\hat{x},\alpha/2)$. It can be verified that (x_n) has a Cauchy subsequence and hence $t(\hat{x}) \ge \frac{\alpha}{2}$ for all $\hat{x} \in X$.

$(c) \Rightarrow (a)$: Since $\hat{\nu}(x) = t(x)$ for all $x \in X$, the implication follows.

$(e) \Rightarrow (g)$: This follows from the fact that $CAT(X) \subseteq AC(\widehat{X})$.

$(g) \Rightarrow (a)$: Let $t(A) > \alpha > 0$ for all $A \in CAT(X)$. Since $x \mapsto \{x\}$ is an isometry, $B(x, \frac{\alpha}{2})$ is totally bounded for all $x \in X$.

$(c) \Rightarrow (h)$: This follows from Theorem 4.11 and Theorem 4.10.

$(h) \Rightarrow (c)$: By usual arguments one can see that $(\widehat{X}, d)$ can be isometrically embedded into the completion of $(AT(X), H_d)$. Since $(\widehat{X}, d)$ is complete, it can be considered as a closed subset of the completion of $(AT(X), H_d)$ and thus $(\widehat{X}, d)$ cofinally complete. Now, apply Theorem 4.11 and Theorem 4.10. □

Remark 4.5. Note that the space $(AC(\widehat{X}), H_d)$ is same as $(CAT(\widehat{X}), H_d)$.

Additionally, one may refer to [Holá and Neubrunn (1988)].

Exercises

Exercise 4.1

Show that the two expressions given for Hausdorff metric H_d on $CL(X)$ in Definition 4.2 are equivalent.

Exercise 4.2

[Beer (1993)] Prove that two metrics on a set X generate the same Hausdorff metric topology on $CL(X)$ if and only if they are uniformly equivalent.

Exercise 4.3

[Beer (1993)] Verify the following relations:

(a) $\tau_{V^+} \subset \tau_V \subseteq \tau_{lf}$,
(b) $\tau_{\delta_d} \subseteq \tau_{H_d} \subseteq \tau_{lf}$ and
(c) $\tau_{\delta_d} \subseteq \tau_V$
(d) $H_\sigma(f,g) \leq d_{uc}(f,g)\ \forall\ f,\ g \in C(X,Y)$.

Exercise 4.4

Prove Proposition 4.2.

Exercise 4.5

Let (X,d) be a metric space. Then prove the following statements to be equivalent:

(a) (X,d) is UC.
(b) (Beer [Beer (1985)]) For any metric space (Y,ρ), $\tau_{H_\sigma} = \tau_{uc}$ on $C(X,Y)$.
(c) (Beer, Himmelberg, Prikry and Vleck [Beer *et al.* (1987)]) $\tau_{\delta_d} = \tau_{lf}$ on $CL(X)$.

Exercise 4.6

Verify that the topology of pointwise convergence is weaker than the proximal topology on $C(X,Y)$.

Exercise 4.7

[Beer (1993)] Show that a net (f_λ) converges to f in $C(X,Y)$ with proximal topology if and only if $\lim_\lambda e_\sigma(G(f_\lambda),G(f)) = 0$, where $e_\sigma(A,B) = \inf\{\varepsilon > 0 : A \subseteq B(B,\varepsilon)\}$.

Exercise 4.8

Prove Theorem 4.1 and Proposition 4.3.

Exercise 4.9

[Beer (1985)] Show that the hyperspace $(CL(X),H_d)$ is a UC space if and only if either X is compact or X is uniformly discrete.

Exercise 4.10

[Beer and Di Maio (2010)] For a metric space (X,d), prove that the following assertions are equivalent.

(a) X is a point of local compactness of $(CL(X),H_d)$.
(b) The hyperspace $(CL(X),H_d)$ is uniformly locally compact.
(c) The hyperspace $(CL(X),H_d)$ is cofinally complete.

Exercise 4.11

[Künzi and Romaguera (1999)] A metric space (X,d) is uniformly locally compact if and only if the set of non-empty compact subsets of X equipped with the Hausdorff metric is cofinally complete.

Exercise 4.12

[Jain and Kundu (2008)] Prove that for a metric space (X,d), the completion of $(CL(X),H_d)$ is $(CL(\widehat{X}),H_d)$.

Exercise 4.13

[Gupta and Kundu (2021a)] For a metric space (X,d), prove that the following assertions are equivalent.

(a) The metric space $(\widehat{X},d)$ is cofinally complete.
(b) On $CAT(X)$, $\tau_{lf} = \tau_{H_d}$.
(c) On $CAT(X)$, $\tau_V \subseteq \tau_{H_d}$.

(d) On $CAT(X)$, $\tau_{V^+} \subseteq \tau_{H_d}$.
(e) On $CAT(X)$, $\tau_V = \tau_{\delta_d}$.
(f) On $CAT(X)$, $\tau_{V^+} \subseteq \tau_{\delta_d}$.

Exercise 4.14

[Gupta and Kundu (2021a)] For a metric space (X,d), prove that the following assertions are equivalent.

(a) The metric space $(\widehat{X},d)$ is cofinally complete.
(b) Whenever (Y,ρ) is a metric space, then $\tau_{\delta_\sigma} = \tau_{uc}$ on $FV(X,Y)$.
(c) On $FV(X)$, $\tau_{\delta_\sigma} = \tau_{uc}$.
(d) There exists a metric space (Y,ρ) containing a nonconstant path for which $\tau_{uc} = \tau_{\delta_\sigma}$ on $FV(X,Y)$.
(e) There exists a metric space (Y,ρ) containing a nonconstant path for which $\tau_{H_\sigma} = \tau_{uc}$ on $FV(X,Y)$.
(f) Whenever (Y,ρ) is a metric space, then $\tau_{H_\sigma} = \tau_{uc}$ on $FV(X,Y)$.
(g) On $FV(X)$, $\tau_{H_\sigma} = \tau_{uc}$.

Chapter 5

Stronger Cofinal Completeness

This chapter is dedicated to the study of stronger versions of cofinally complete metric spaces, namely UC spaces and cofinally Bourbaki-complete spaces. Several characterizations of UC spaces and the spaces having UC completion are studied in terms of relations among various functions such as Lipschitz-type functions, continuous functions, CC-regular functions, PC-regular functions and Cauchy-subregular functions. Furthermore, we investigate certain properties of CBC-regular functions which lead us to interesting characterizations of cofinally Bourbaki-complete metric spaces and finitely chainable spaces.

5.1 UC Spaces

In Chapter 2, we have seen that every uniformly continuous function is 'almost uniformly continuous', that is, every uniformly continuous function is both continuous and CC-regular. Also, we know that every uniformly continuous function between two metric spaces is continuous and has property P, where being P is any one of PC-regular, uniformly locally bounded and Cauchy-subregular. We now look for the conditions under which the converse is true, that is, the conditions under which continuity and the property P (as stated before) ensures uniform continuity.

Theorem 5.1. *Let (X,d) be a metric space. Then the following statements are equivalent.*

- (a) *(X,d) is a UC space.*
- (b) *Whenever (Y,ρ) is a metric space and $f:(X,d)\to(Y,\rho)$ is both continuous and CC-regular, then f is uniformly continuous.*
- (c) *Whenever (Y,ρ) is a metric space and $f:(X,d)\to(Y,\rho)$ is both locally Lipschitz and CC-regular, then f is uniformly continuous.*

(*d*) *If* $f : (X,d) \to \mathbb{R}$ *is both locally Lipschitz and CC-regular, then* f *is uniformly continuous.*

(*e*) *Whenever* (Y,ρ) *is a metric space and* $f : (X,d) \to (Y,\rho)$ *is both continuous and PC-regular, then* f *is uniformly continuous.*

(*f*) *Whenever* (Y,ρ) *is a metric space and* $f : (X,d) \to (Y,\rho)$ *is both locally Lipschitz and PC-regular, then* f *is uniformly continuous.*

(*g*) *If* $f : (X,d) \to \mathbb{R}$ *is both locally Lipschitz and PC-regular, then* f *is uniformly continuous.*

(*h*) *Whenever* (Y,ρ) *is a metric space and* $f : (X,d) \to (Y,\rho)$ *is both continuous and uniformly locally bounded, then* f *is uniformly continuous.*

(*i*) *Whenever* (Y,ρ) *is a metric space and* $f : (X,d) \to (Y,\rho)$ *is both locally Lipschitz and uniformly locally bounded, then* f *is uniformly continuous.*

(*j*) *If* $f : (X,d) \to \mathbb{R}$ *is both locally Lipschitz and uniformly locally bounded, then* f *is uniformly continuous.*

(*k*) *Whenever* (Y,ρ) *is a metric space and* $f : (X,d) \to (Y,\rho)$ *is both continuous and Cauchy-subregular, then* f *is uniformly continuous.*

(*l*) *Whenever* (Y,ρ) *is a metric space and* $f : (X,d) \to (Y,\rho)$ *is both locally Lipschitz and Cauchy-subregular, then* f *is uniformly continuous.*

(*m*) *If* $f : (X,d) \to \mathbb{R}$ *is both locally Lipschitz and Cauchy-subregular, then* f *is uniformly continuous.*

Proof. The statements $(a) \Rightarrow (b) \Rightarrow (c) \Rightarrow (d)$, $(a) \Rightarrow (e) \Rightarrow (f) \Rightarrow (g)$, $(a) \Rightarrow (h) \Rightarrow (i) \Rightarrow (j)$ and $(a) \Rightarrow (k) \Rightarrow (l) \Rightarrow (m)$ are all immediate.

$(d) \Rightarrow (a)$: First we prove that (X,d) is complete. Suppose (x_n) is a Cauchy sequence of distinct points in (X,d) such that it does not converge. Thus the set $A = \{x_n : n \in \mathbb{N}\}$ is closed and discrete. Consequently, $\forall n \in \mathbb{N}$, $\exists \varepsilon_n > 0$ such that $d(x_m,x_n) > \varepsilon_n \ \forall m \neq n$. Let $\delta_n = \min\{1/n, \varepsilon_n/3\}$. Define a function $f : (X,d) \to [0,2]$ as follows:

$$f(x) = \begin{cases} 1 - \frac{1}{\delta_n} d(x,x_n) & : x \in B(x_n,\delta_n) \text{ for some odd } n \in \mathbb{N} \\ 2 - \frac{2}{\delta_n} d(x,x_n) & : x \in B(x_n,\delta_n) \text{ for some even } n \in \mathbb{N} \\ 0 & : \text{otherwise} \end{cases}$$

Clearly, f restricted to each ball $B(x_n,\delta_n)$ is Lipschitz. Let $x \in X$, since x is not a cluster point of the sequence (x_n) and $\varepsilon_n \leq 1/n \ \forall n \in \mathbb{N}$, $\exists \delta_x > 0$ such that $B(x,\delta_x)$ intersects at most one of the balls $B(x_n,\delta_n)$. Thus f is locally Lipschitz. Since the range of f is totally bounded, f is CC-regular as well, but f is not uniformly continuous since every uniformly continuous functions is Cauchy-regular. We get a contradiction. Hence (X,d) is complete.

To see that (X,d) is a UC space, let (x_n) be a sequence of distinct points in X such that $\lim_{n\to\infty} I(x_n) = 0$. Suppose the sequence does not cluster. Since (X,d) is complete, (x_n) has no Cauchy subsequence. Thus $\exists\, \delta > 0$ such that by passing to a subsequence, if needed, the family of open balls $\{B(x_n, \delta) : n \in \mathbb{N}\}$ is pairwise disjoint. Also, $\lim_{n\to\infty} I(x_n) = 0$ implies that we can assume $I(x_n) < \min\{\delta, \frac{1}{n}\}\ \forall n \in \mathbb{N}$. Let $\delta_n = \min\{\delta, \frac{1}{n}\}\ \forall n \in \mathbb{N}$. Consequently, $\forall n \in \mathbb{N}$, we can choose $y_n \neq x_n$ such that $d(x_n, y_n) < \delta_n$. Let $\varepsilon_n = d(x_n, y_n)\ \forall n \in \mathbb{N}$. Define a function $f : (X,d) \to [0,1]$ as follows:

$$f(x) = \begin{cases} 1 - \frac{1}{\varepsilon_n} d(x, x_n) & : x \in B(x_n, \varepsilon_n) \text{ for some } n \in \mathbb{N} \\ 0 & : \text{otherwise} \end{cases}$$

Note that f is locally Lipschitz and CC-regular but not uniformly continuous as $d(x_n, y_n)$ tends to 0 but $|f(x_n) - f(y_n)| = 1\ \forall n \in \mathbb{N}$. We get a contradiction. Thus by Proposition 1.3 (X,d) is a UC space.

In a manner similar to the proof of $(d) \Rightarrow (a)$, the implications $(g) \Rightarrow (a)$ and $(j) \Rightarrow (a)$ can be proved. The implication $(m) \Rightarrow (g)$ follows from Proposition 1.7 and Proposition 3.3. □

Remark 5.1.

- (i) The previous result was proved in [Gupta and Kundu (2020)], except for the equivalence of (k), (l) and (m) with (a) which was proved in [Gupta and Kundu (2022)].
- (ii) It is useful to note that Theorem 5.1 characterizes UC-ness of a metric space in terms of the uniform continuity of various thin subclasses of continuous functions.

It is known that every real-valued Cauchy-regular function on a metric space (X,d) is uniformly continuous if and only if $(\widehat{X},d)$ is UC [Beer (1986); Jain and Kundu (2007)]. Thus it is always useful to study the cases for which the completion of a metric space is UC or cofinally complete. Moreover, since these spaces are complete, it is natural to consider the metric spaces which have completions that are UC, cofinally complete, cofinally Bourbaki-complete etc. In our next theorem, we characterize the metric spaces having UC completion in terms of uniform continuity of several thin subclasses of Cauchy-regular functions. But before that, we would like to make a small observation: if we consider the function f given in Example 2.3 with the usual metric on the set Y, the function would be Cauchy-Lipschitz but not uniformly locally bounded. Thus a Cauchy-Lipschitz function between two metric spaces need not be uniformly locally bounded.

Proposition 5.1. *[Beer (1986); Jain and Kundu (2007)] The completion* $(\widehat{X},d)$ *of a metric space* (X,d) *is UC if and only if for every sequence* (x_n) *in* X *with* $\lim_{n\to\infty} I(x_n) = 0$ *has a Cauchy subsequence.*

Theorem 5.2. *[Gupta and Kundu (2020)] Let* (X,d) *be a metric space. Then the following assertions are equivalent.*

- (a) *The completion* $(\widehat{X},d)$ *of* (X,d) *is a UC space.*
- (b) *Whenever* (Y,ρ) *is a metric space and* $f:(X,d)\to(Y,\rho)$ *is both Cauchy-regular and CC-regular, then* f *is uniformly continuous.*
- (c) *Whenever* (Y,ρ) *is a metric space and* $f:(X,d)\to(Y,\rho)$ *is both Cauchy-Lipschitz and CC-regular, then* f *is uniformly continuous.*
- (d) *Whenever* (Y,ρ) *is a metric space and* $f:(X,d)\to(Y,\rho)$ *is both uniformly locally Lipschitz and CC-regular, then* f *is uniformly continuous.*
- (e) *If* $f:(X,d)\to\mathbb{R}$ *is uniformly locally Lipschitz, then* f *is uniformly continuous.*
- (f) *Whenever* (Y,ρ) *is a metric space and* $f:(X,d)\to(Y,\rho)$ *is both Cauchy-regular and PC-regular, then* f *is uniformly continuous.*
- (g) *Whenever* (Y,ρ) *is a metric space and* $f:(X,d)\to(Y,\rho)$ *is both Cauchy-Lipschitz and PC-regular, then* f *is uniformly continuous.*
- (h) *Whenever* (Y,ρ) *is a metric space and* $f:(X,d)\to(Y,\rho)$ *is both uniformly locally Lipschitz and PC-regular, then* f *is uniformly continuous.*
- (i) *If* $f:(X,d)\to\mathbb{R}$ *is both uniformly locally Lipschitz and PC-regular, then* f *is uniformly continuous.*
- (j) *Whenever* (Y,ρ) *is a metric space and* $f:(X,d)\to(Y,\rho)$ *is both Cauchy-regular and uniformly locally bounded, then* f *is uniformly continuous.*
- (k) *Whenever* (Y,ρ) *is a metric space and* $f:(X,d)\to(Y,\rho)$ *is both Cauchy-Lipschitz and uniformly locally bounded, then* f *is uniformly continuous.*
- (l) *If* $f:(X,d)\to\mathbb{R}$ *is both Cauchy-Lipschitz and uniformly locally bounded, then* f *is uniformly continuous.*

Proof. The implications $(b)\Rightarrow(c)\Rightarrow(d)$, $(f)\Rightarrow(g)\Rightarrow(h)\Rightarrow(i)$, $(j)\Rightarrow(k)\Rightarrow(l)$ are all immediate.

$(a)\Rightarrow(b)$, $(a)\Rightarrow(f)$ and $(a)\Rightarrow(j)$: Let $f:(X,d)\to(Y,\rho)$ be a Cauchy-regular function. By Proposition 1.8, there exists a Cauchy-regular function $\hat{f}:(\widehat{X},d)\to(\widehat{Y},\rho)$ such that $\hat{f}|_X = f$. Since $(\widehat{X},d)$ is UC, $\hat{f}$ is uniformly continuous and hence f is uniformly continuous.

$(d)\Rightarrow(e)$: Since every uniformly locally Lipschitz function is uniformly locally bounded and every real-valued uniformly locally bounded function is CC-regular (by Proposition 2.1), the implication follows.

$(e) \Rightarrow (a)$: Suppose $(\widehat{X},d)$ is not a UC space. By Proposition 5.1, there exists a sequence of distinct points (x_n) in (X,d) with $\lim_{n\to\infty} I(x_n) = 0$ such that (x_n) does not have any Cauchy subsequence. Thus $\exists\, \delta > 0$ such that by passing to a subsequence, if needed, the family of open balls $\{B(x_n,\delta) : n \in \mathbb{N}\}$ is pairwise disjoint. Let $\delta_n = \min\{\frac{\delta}{4}, \frac{1}{n}\}$ $\forall n \in \mathbb{N}$. Also, $\lim_{n\to\infty} I(x_n) = 0$ implies that we can assume $I(x_n) < \delta_n$ $\forall n \in \mathbb{N}$. Consequently, $\forall n \in \mathbb{N}$, we can choose $y_n \neq x_n$ such that $d(x_n,y_n) < \delta_n$. Let $\varepsilon_n = d(x_n,y_n)$ $\forall n \in \mathbb{N}$. Define a function $f : (X,d) \to [0,1]$ as follows.

$$f(x) = \begin{cases} 1 - \frac{1}{\varepsilon_n} d(x,x_n) & : x \in B(x_n,\varepsilon_n) \text{ for some } n \in \mathbb{N} \\ 0 & : \text{otherwise} \end{cases}$$

The function f is uniformly locally Lipschitz because $\forall x \in X$, $B(x,\frac{\delta}{4})$ intersects at most one of the balls $B(x_m,\frac{\delta}{4})$ and f restricted to each ball $B(x_m,\frac{\delta}{4})$ is Lipschitz. Thus f is uniformly locally Lipschitz but not uniformly continuous, a contradiction.

$(i) \Rightarrow (a)$: Note that the function f defined in the proof of $(e) \Rightarrow (a)$ is PC-regular as well as the set $[0,1]$ is totally bounded in $\mathbb{R}$. Thus in a manner similar to the proof of $(e) \Rightarrow (a)$, $(i) \Rightarrow (a)$ can be proved.

$(l) \Rightarrow (a)$: Let $f : (\widehat{X},d) \to \mathbb{R}$ be a locally Lipschitz uniformly locally bounded function. Since $(\widehat{X},d)$ is a complete metric space, by Theorem 1.3, f is Cauchy-Lipschitz. This implies $f|_X$ is Cauchy-Lipschitz and uniformly locally bounded and hence uniformly continuous. We claim that $f : (\widehat{X},d) \to \mathbb{R}$ is uniformly continuous as well. For $\varepsilon > 0$, $\exists\, \delta > 0$ such that $\forall x,y \in X$ with $d(x,y) < \delta$, $|f(x) - f(y)| < \varepsilon/3$. Let $\hat{x},\hat{y} \in \widehat{X}$ such that $d(\hat{x},\hat{y}) < \delta/4$. Since $\hat{x},\hat{y} \in \widehat{X}$, there exists sequences (x_n) and (y_n) in X such that (x_n) converges to $\hat{x}$ and (y_n) converges to $\hat{y}$. Thus, there exists $n_0 \in \mathbb{N}$ such that $d(x_n,\hat{x}) < \delta/2$, $d(y_n,\hat{y}) < \delta/4$, $|f(\hat{x}) - f(x_n)| < \varepsilon/3$ and $|f(y_n) - f(\hat{y})| < \varepsilon/3$ $\forall n \geq n_0$. Thus for all $n \geq n_0$,

$$d(x_n,y_n) \leq d(x_n,\hat{x}) + d(\hat{x},\hat{y}) + d(y_n,\hat{y}) < \delta$$

Therefore,

$$\begin{aligned} |f(\hat{x}) - f(\hat{y})| &\leq |f(\hat{x}) - f(x_{n_0})| + |f(x_{n_0}) - f(y_{n_0})| + |f(y_{n_0}) - f(\hat{y})| \\ &< \varepsilon \end{aligned}$$

Thus by Theorem 5.1, $(\widehat{X},d)$ is a UC space. □

Moving on to the next set of characterizations of UC spaces, let us recall that a continuous function between two metric spaces need not be PC-regular. For example, consider the function $f : \mathbb{R} \to \mathbb{R}$ defined as : $f(x) = x^2$. Then f is in fact Cauchy-regular, but f is not PC-regular as the image of the pseudo-Cauchy sequence $(1,\ 1+1,\ 2,\ 2+\frac{1}{2},\ 3,\ 3+\frac{1}{3},\ldots)$ is not pseudo-Cauchy under f. Clearly, if (X,d) is a UC space then every real-valued continuous function on (X,d) is PC-regular. The next result says that the converse is true as well. Before stating the result, we would like to generalize the notion of asymptotic sequences.

Definitions 5.1. A pair of sequences (x_n) and (y_n) in a metric space (X,d) is said to be:

(a) *uniformly asymptotic*, written $(x_n) \asymp^u (y_n)$, if $\forall\, \varepsilon > 0$, $\exists\, n_o \in \mathbb{N}$ such that $d(x_m, y_n) < \varepsilon\ \forall\, m,\ n > n_o$.
(b) *cofinally asymptotic*, written $(x_n) \asymp_c (y_n)$, if $\forall\, \varepsilon > 0$, $\exists$ an infinite subset N_ε of $\mathbb{N}$ such that $d(x_n, y_n) < \varepsilon\ \forall\, n \in N_\varepsilon$.
(c) *Bourbaki-asymptotic* in (X,d), written $(x_n) \asymp_b (y_n)$, if $\forall\, \varepsilon > 0$, $\exists\, m,\ n_o \in \mathbb{N}$ such that x_n and y_n can be joined by an ε-chain of length m for all $n \geq n_o$.
(d) *cofinally Bourbaki-asymptotic* in X, written $(x_n) \asymp_{cb} (y_n)$, if $\forall\, \varepsilon > 0$, $\exists\, m \in \mathbb{N}$ and an infinite subset N_ε of $\mathbb{N}$ such that x_n and y_n can be joined by an ε-chain of length m for all $n \in N_\varepsilon$.

Note that in [Snipes (1977)] asymptotic sequences and uniformly asymptotic sequences in a metric space were called parallel sequences and equivalent sequences respectively, while in [Borsík (2000)] such sequences were considered in the context of uniform spaces. We have already mentioned in Section 3.2, how asymptotic sequences characterize uniformly continuous functions. The next proposition says that uniformly continuous functions can be characterized by pairs of cofinally asymptotic sequences as well. We omit the routine proof.

Proposition 5.2. *Let $f : (X,d) \to (Y,\rho)$ be a function between two metric spaces. Then the following statements are equivalent.*

(a) *f is uniformly continuous.*
(b) *Whenever $(x_n) \asymp_c (z_n)$ in (X,d) then $(f(x_n)) \asymp_c (f(z_n))$ in (Y,ρ).*
(c) *Whenever $(x_n) \asymp (z_n)$ in (X,d) then $(f(x_n)) \asymp_c (f(z_n))$ in (Y,ρ).*

Theorem 5.3. *Let (X,d) be a metric space. Then the following statements are equivalent.*

(a) *(X,d) is a UC space.*
(b) *Whenever (x_n) and (y_n) is a pair of cofinally asymptotic sequences in (X,d) such that $x_n \neq y_n$ for all $n \in \mathbb{N}$, then both (x_n) and (y_n) cluster in X.*
(c) *Every continuous function on (X,d) with values in a metric space (Y,ρ) preserves pairs of cofinally asymptotic sequences.*

(*d*) *Every continuous function on* (X,d) *with values in a metric space* (Y,ρ) *is PC-regular.*
(*e*) *Every locally Lipschitz function on* (X,d) *with values in a metric space* (Y,ρ) *is PC-regular.*
(*f*) *Every locally Lipschitz function on* (X,d) *with values in a metric space* (Y,ρ) *is uniformly continuous.*
(*g*) *Every locally Lipschitz function on* (X,d) *with values in a metric space* (Y,ρ) *preserves pairs of cofinally asymptotic sequences.*
(*h*) *For every metric* σ *equivalent to d and every* $\varepsilon > 0$*, the open cover* $\mathscr{V} = \{B_\sigma(x,\varepsilon) : x \in X\}$ *has a Lebesgue number with respect to d.*
(*i*) *For every metric* σ *equivalent to d and every* $\varepsilon > 0$*, there exists* $\delta > 0$ *such that for all* $x \in X,\ B_d(x,\delta) \subseteq B_\sigma(x,\varepsilon)$.
(*j*) *Every d-pseudo-Cauchy sequence in X is* σ*-pseudo-Cauchy for all equivalent metrics* σ *on X.*

Proof. $(a) \Rightarrow (b)$: Let $(x_n) \asymp_c (y_n)$ such that $x_n \neq y_n\ \forall\ n \in \mathbb{N}$. It can be verified that there exist subsequences $(x_{n_k})_{k\in\mathbb{N}}$ and $(y_{n_k})_{k\in\mathbb{N}}$ of (x_n) and (y_n) respectively, such that $(x_{n_k}) \asymp (y_{n_k})$. Thus by Proposition 1.3, the result follows.

$(b) \Rightarrow (c)$: Let $f : (X,d) \to (Y,\rho)$ be a continuous function. We prove that under the given conditions, the function f not just preserves pairs of cofinally asymptotic sequences but even it is uniformly continuous. Suppose f is not uniformly continuous. Then there exists an $\varepsilon_o > 0$ such that for all $n \in \mathbb{N},\ \exists\ x_n,\ z_n \in X$ with $d(x_n,z_n) < \frac{1}{n}$ and $\rho(f(x_n),f(z_n)) > \varepsilon_o$. Thus, $(x_n) \asymp (z_n)$ and $x_n \neq z_n\ \forall\ n \in \mathbb{N}$. Let $(x_{n_k})_{k\in\mathbb{N}}$ and $(z_{n_k})_{k\in\mathbb{N}}$ be subsequences of (x_n) and (z_n) respectively which converge to some $x \in X$. Then the sequence (v_n) defined as: $v_{2k-1} = x_{n_k}$ and $v_{2k} = z_{n_k}$ $\forall\ k \in \mathbb{N}$, converges to x. Hence by the continuity of f, $(f(v_n))$ converges to $f(x)$. We get a contradiction. Thus, f is uniformly continuous.

$(c) \Rightarrow (d)$: Let $f : (X,d) \to (Y,\rho)$ be a continuous function and let (x_n) be a pseudo-Cauchy sequence in (X,d). By passing to a subsequence, we can assume that $d(x_{2n-1},x_{2n}) < \frac{1}{n}\ \forall\ n \in \mathbb{N}$. Suppose $(f(x_n))$ is not pseudo-Cauchy in (Y,ρ). Then there exist $\varepsilon_o > 0$ and $n_o \in \mathbb{N}$ such that $\rho(f(x_n),f(x_m)) > \varepsilon_o\ \ \forall\ n,m > n_o,\ n \neq m$. Since $(x_{2n-1}) \asymp_c (x_{2n})$, $(f(x_{2n-1})) \asymp_c (f(x_{2n}))$. Thus there exists an infinite subset N of $\mathbb{N}$ such that for every $n \in N$, $\rho(f(x_{2n-1}),f(x_{2n})) < \varepsilon_o$. We arrive at a contradiction. Thus f is PC-regular.

$(d) \Rightarrow (e)$: This is immediate.

$(e) \Rightarrow (f)$: Let $f : (X,d) \to (Y,\rho)$ be a locally Lipschitz function which is not uniformly continuous. Then there exists an $\varepsilon_o > 0$ such that for all $n \in \mathbb{N}$, $\exists\ x_n,\ y_n \in X$ with $d(x_n,y_n) < \frac{1}{n}$ and $\rho(f(x_n),f(y_n)) > \varepsilon_o$. We claim that the sequence (x_n) has no cluster point. Suppose it has a cluster point, say x. Then there exists a subsequence $(x_{m_k})_{k\in\mathbb{N}}$ of (x_n) converging to x. Subsequently, the subsequence $(y_{m_k})_{k\in\mathbb{N}}$ of (y_n) converges to x. By the continuity of f, the sequences $(f(x_{m_k}))$ and $(f(y_{m_k}))$ converges to $f(x)$. Thus, eventually $\rho(f(x_{m_k}),f(y_{m_k})) \leq \rho(f(x_{m_k}),f(x))+\rho(f(x),f(y_{m_k})) \leq \varepsilon_o$. We arrive at a contradiction. Hence the sequences (x_n) and (y_n) have no cluster point.

We claim that there exists a strictly increasing sequence (n_k) in $\mathbb{N}$ such that the subsequences (x_{n_k}) and (y_{n_k}) of (x_n) and (y_n) respectively satisfy the following: for all $k > 1$, $x_{n_k},\ y_{n_k} \notin \{x_{n_1},\dots,x_{n_{k-1}},y_{n_1},\dots,y_{n_{k-1}}\}$. We prove it by induction. For $k = 1$, choose $n_1 = 1$. Suppose for $k = 2,\dots,m$, we have chosen $n_2,\dots,n_m$ such that $n_1 < n_2 < \cdots < n_m$ and $x_{n_k},\ y_{n_k} \notin \{x_{n_1},\dots,x_{n_{k-1}},y_{n_1},\dots,y_{n_{k-1}}\}$ for $1 < k \leq m$. Suppose there does not exist any n_{m+1} with the required property. Thus, for every $n > n_m$ either $x_n \in \{x_{n_1},\dots,x_{n_m},y_{n_1},\dots,y_{n_m}\}$ or $y_n \in \{x_{n_1},\dots,x_{n_m},y_{n_1},\dots,y_{n_m}\}$. In either case we will get a contradiction as the sequences (x_n) and (y_n) have no cluster point. Hence the claim is proved.

Now let $A = \{x_{n_k},\ y_{n_k} : k \in \mathbb{N}\}$, Then A is closed and discrete and hence the function $g : (A,d) \to \mathbb{R}$ defined as: $g(x_{n_k}) = 2k,\ g(y_{n_k}) = 2k+1$, is continuous. By Tietze's extension theorem, there exists a real-valued continuous extension of g to (X,d), say g'. Then by Theorem 1.4, for $0 < \varepsilon < 1$, there exists a locally Lipschitz function $h : (X,d) \to \mathbb{R}$ with $\sup_{x\in X} |g'(x)-h(x)| \leq \frac{\varepsilon}{3}$. Thus by (e), h is PC-regular which implies that the image of the sequence (v_n) defined as: $v_{2k-1} = x_{n_k}$ and $v_{2k} = y_{n_k}\ \ \forall\, k \in \mathbb{N}$, under h is pseudo-Cauchy. Consequently, for every $n \in \mathbb{N}$ there exist $k > m > n$ such that $|h(v_k)-h(v_m)| < \frac{\varepsilon}{3}$. Hence $1 \leq |k-m| = |g(v_k)-g(v_m)| \leq |g(v_k)-h(v_k)|+|h(v_k)-h(v_m)|+|h(v_m)-g(v_m)| < \varepsilon < 1$. We arrive at a contradiction. Thus f is uniformly continuous.

$(f) \Leftrightarrow (g)$: This follows from Proposition 5.2.

$(f) \Rightarrow (a)$: This follows from Theorem 5.1

$(a) \Rightarrow (h)$: Suppose $\mathscr{V}$ has no Lebesgue number with respect to d. Then for each $n \in \mathbb{N}$, there exists $x_n \in X$ such that $B_d(x_n,\frac{1}{n})$ is contained in no single member of $\mathscr{V}$. Thus for each $n \in \mathbb{N}$, $I(x_n) < \frac{1}{n}$ and hence by Proposition 1.3 the sequence (x_n) has a cluster point say a. Then $a \in V$ for some $V \in \mathscr{V}$. Since V is open, it contains infinitely many terms of the sequence. We get a contradiction.

$(h) \Rightarrow (i)$: Let $\delta > 0$ be a Lebesgue number for the open cover $\mathscr{V} = \{B_\sigma(x,\frac{\varepsilon}{2}) : x \in X\}$ with respect to d. Fix $x \in X$ and choose $a \in X$ such that $B_d(x,\delta) \subseteq B_\sigma(a,\frac{\varepsilon}{2})$. This implies that $B_d(x,\delta) \subseteq B_\sigma(x,\varepsilon)$.

$(i) \Rightarrow (j)$: Let (x_n) be a d-pseudo-Cauchy sequence in X and let $\varepsilon > 0$. Then there exists $\delta > 0$ such that for all $x \in X$, $B_d(x,\delta) \subseteq B_\sigma(x,\varepsilon)$. Now for this δ and for all $n \in \mathbb{N}$, there exist $j,\ k > n\ \ j \neq k$ such that $d(x_j,x_k) < \delta$. Consequently, $x_j \in B_d(x_k,\delta) \subseteq B_\sigma(x_k,\varepsilon)$ and hence $\sigma(x_j,x_k) < \varepsilon$. Thus, (x_n) is a σ-pseudo-Cauchy sequence.

$(j) \Rightarrow (d)$: Let $f : (X,d) \to (Y,\rho)$ be a continuous function. Then the metric σ defined by,

$$\sigma(a,b) = d(a,b) + \rho(f(a), f(b)) \text{ for } a,\ b \text{ in } X$$

is equivalent to d. Now if (x_n) is a d-pseudo-Cauchy sequence in X, then it is σ-pseudo-Cauchy and hence $(f(x_n))$ is ρ-pseudo-Cauchy in Y. □

Remark 5.2. (i) Theorem 5.3 still holds if we replace the metric space (Y,ρ) by $(\mathbb{R}, |\cdot|)$.
(ii) The equivalence of conditions (a), (h) and (i) was proved in [Beer and Di Maio (2012)].

In Proposition 1.7, we have already seen that every PC-regular function is CC-regular. Now we look into the conditions on a metric space (X,d) under which every real-valued CC-regular function on (X,d) is PC-regular and we observe that such metric spaces are precisely those spaces whose completions are UC. But before proving that we would like to mention a result from [Jain and Kundu (2007)] without proof, which is actually a small observation regarding uniformly asymptotic sequences.

Lemma 5.1. *Let (x_n) and (y_n) be two sequences in a metric space (X,d). Then $(x_n) \asymp^u (y_n)$ if and only if $(x_n) \asymp (y_n)$ and (x_n) is Cauchy in (X,d).*

Theorem 5.4. *Let (X,d) be a metric space. Then the following statements are equivalent.*

- (a) *The completion $(\widehat{X},d)$ of (X,d) is a UC space.*
- (b) *Whenever $(x_n) \asymp_c (y_n)$ such that $x_n \neq y_n$ for each $n \in \mathbb{N}$, then there exists a strictly increasing sequence (n_k) in $\mathbb{N}$ such that $(x_{n_k}) \asymp^u (y_{n_k})$.*
- (c) *Every sequence in X with distinct terms is either uniformly discrete or else it has a Cauchy subsequence.*
- (d) *Every sequence in X with distinct terms is either uniformly discrete or else cofinally Cauchy.*
- (e) *Every pseudo-Cauchy sequence in (X,d) with distinct terms is cofinally Cauchy.*
- (f) *Every CC-regular function on (X,d) with values in a metric space (Y,ρ) is PC-regular.*
- (g) *Every Cauchy-Lipschitz function on (X,d) with values in an arbitrary metric space (Y,ρ) is Lipschitz in the small.*
- (h) *Every Cauchy-Lipschitz function on (X,d) with values in a metric space (Y,ρ) is uniformly continuous.*

(i) *Every Cauchy-Lipschitz function on (X,d) with values in a metric space (Y,ρ) is PC-regular.*

(j) *Every Cauchy-regular function on (X,d) with values in a metric space (Y,ρ) is PC-regular.*

(k) *Every d-pseudo-Cauchy sequence in X is σ-pseudo-Cauchy for all Cauchy equivalent metrics σ on X.*

Proof. The implications $(c) \Rightarrow (d) \Rightarrow (e)$, $(g) \Rightarrow (h) \Rightarrow (i)$ and $(j) \Rightarrow (k)$ are immediate.

$(a) \Rightarrow (b)$: Let $(x_n) \asymp_c (y_n)$ such that $x_n \neq y_n \; \forall \; n \in \mathbb{N}$. Again, there exists a strictly increasing sequence (n_k) in $\mathbb{N}$ such that $(x_{n_k}) \asymp (y_{n_k})$. Now the result follows from Proposition 5.1.

$(b) \Rightarrow (c)$: Suppose (x_n) is a sequence in X with distinct terms, which is not uniformly discrete. Then there exists a subsequence $(x_{n_k})_{k\in\mathbb{N}}$ of (x_n) such that $d(x_{n_{2k}}, x_{n_{2k+1}}) < \frac{1}{k}$ for all $k \in \mathbb{N}$. Hence $(x_{n_{2k}}) \asymp_c (x_{n_{2k+1}})$. Using (b) and Lemma 5.1, we get a Cauchy subsequence of (x_n).

$(e) \Rightarrow (f)$: Let $f : (X,d) \to (Y,\rho)$ be a CC-regular function and let (x_n) be a pseudo-Cauchy sequence in (X,d). If (x_n) has a pseudo-Cauchy subsequence of distinct terms or if it has a constant subsequence, then we are done; otherwise the range of the function, $g(n) = x_n$ for $n \in \mathbb{N}$, has infinitely many terms which have more than one pre-image and hence $(f(x_n))$ is pseudo-Cauchy in (Y,ρ).

$(f) \Rightarrow (e)$: Let (x_n) be a pseudo-Cauchy sequence in (X,d) with distinct terms which is not cofinally Cauchy. Then the function $f : (X,d) \to \mathbb{R}$ defined by:

$$f(x) = \left\{ \begin{array}{l} n : x = x_n \text{ for some } n \in \mathbb{N} \\ 0 : \text{otherwise} \end{array} \right\}$$

is CC-regular but not PC-regular, which is a contradiction.

$(e) \Rightarrow (a)$: Suppose $(\widehat{X},d)$ is not UC; then there exists a pseudo-Cauchy sequence (x_n) in (X,d) with distinct terms which has no Cauchy subsequence. Thus no infinite subset of $A = \{x_n : n \in \mathbb{N}\}$ is totally bounded and hence every subsequence of (x_n) contains a uniformly discrete subsequence. Since (x_n) is cofinally Cauchy with no Cauchy subsequence, as in the proof of Proposition 1.1, there exists a pairwise disjoint family $\{\mathbb{M}_j : j \in \mathbb{N}\}$ of infinite subsets of $\mathbb{N}$ such that: (1) if $\{i,\ l\} \subseteq \bigcup\{\mathbb{M}_j : j \in \mathbb{N}\}$ then $x_i \neq x_l$; and (2) if $i,\ l \in \mathbb{M}_j$ then $\varepsilon_j < d(x_i, x_l) < \varepsilon_{j-1}$ where $0 < \varepsilon_j < \frac{1}{j+1}$ and $\varepsilon_o = 1$. Let $(x_{j_k})_{k\in\mathbb{N}}$ be a subsequence of (x_n) such that $j_k \in \mathbb{M}_k \; \forall \; k \in \mathbb{N}$. Then (x_{j_k}) has a uniformly discrete subsequence, for convenience we call it $(x_{j_k})_{k\in\mathbb{N}}$ itself, which implies that there exists $c > 0$ such that $d(x_{j_k}, x_{j_l}) > c \; \forall \; k \neq l$. Now consider the sequence $(a_n) = (x_{j_1}, x_{m_1}, x_{j_2}, x_{m_2}, \ldots)$ where $(x_{m_k})_{k\in\mathbb{N}}$ is another subsequence of (x_n) such that $m_k \in \mathbb{M}_k$ and $m_k \neq j_k$. Then (a_n) is pseudo-Cauchy with distinct terms and hence cofinally Cauchy. Since for every $k \in \mathbb{N}$, $d(x_{j_k}, x_{m_k}) < \frac{1}{k}$, there exists $n_o \in \mathbb{N}$ such that $d(x_{m_k}, x_{m_l}) > \frac{c}{3}\ \forall\ k,\ l \geq n_o,\ k \neq l$. Hence there exists $c_1 > 0$ such that $d(x_{j_k}, x_{j_l}) > c_1$ and $d(x_{m_k}, x_{m_l}) > c_1\ \forall\ k \neq l$. But this is a contradiction as (a_n) is cofinally Cauchy. Thus (x_n) has a Cauchy subsequence.

$(a) \Rightarrow (g)$: Let $f : (X,d) \to (Y,\rho)$ be a Cauchy-Lipschitz function which is not Lipschitz in the small. Thus for each $n \in \mathbb{N}$, there exist x_n and w_n in X with $0 < d(x_n, w_n) < \frac{1}{n}$ and $\rho(f(x_n), f(w_n)) > nd(x_n, w_n)$. Since $I(x_n) < \frac{1}{n}$, by Proposition 5.1, (x_n) has a Cauchy subsequence, say $(x_{n_k})_{k \in \mathbb{N}}$. Then $(x_{n_1}, w_{n_1}, x_{n_2}, w_{n_2}, \ldots)$ is a Cauchy sequence on which f fails to be Lipschitz, a contradiction.

$(i) \Rightarrow (j)$: Let $f : (X,d) \to (Y,\rho)$ be a Cauchy-regular function and let (x_n) be a pseudo-Cauchy sequence of distinct points in (X,d). Suppose $(f(x_n))$ is not pseudo-Cauchy in (Y,ρ). Then (x_n) has no Cauchy subsequence. Consider the function $f : (A,d) \to \mathbb{R}$: $f(x_n) = n$, where $A = \{x_n : n \in \mathbb{N}\}$. Since any Cauchy sequence in (A,d) is eventually constant, the function f is Cauchy-regular. By Corollary 1.1, f can be extended to a real-valued Cauchy-regular function f' on (X,d). Hence by Theorem 1.5, for $0 < \varepsilon < 1$, there exists a Cauchy-Lipschitz function $g : (X,d) \to \mathbb{R}$ with $\sup_{x \in X} |f'(x) - g(x)| \leq \frac{\varepsilon}{3}$. By (i), g is PC-regular and hence $(g(x_n))$ is a pseudo-Cauchy sequence in $\mathbb{R}$. Consequently, for every $n \in \mathbb{N}$, there exist $k > m > n$ such that $|g(x_k) - g(x_m)| < \frac{\varepsilon}{3}$. Furthermore, $1 \leq |k - m| = |f(x_k) - f(x_m)| \leq |f(x_k) - g(x_k)| + |g(x_k) - g(x_m)| + |g(x_m) - f(x_m)| < \varepsilon < 1$, a contradiction. Thus $(f(x_n))$ is pseudo-Cauchy in (Y,ρ).

$(k) \Rightarrow (a)$: Let (x_n) be a pseudo-Cauchy sequence of distinct points in (X,d). Suppose that (x_n) has no Cauchy subsequence. Then as in the proof of the implication $(i) \Rightarrow (j)$, there exists a real-valued Cauchy-regular function f' on (X,d) which is not PC-regular. Now consider the metric σ on X,

$$\sigma(a,b) = d(a,b) + |f'(a) - f'(b)| \text{ for } a,\ b \text{ in } X$$

Then σ is Cauchy-equivalent to d, but (x_n) is not σ-pseudo-Cauchy as $\sigma(x_n, x_m) \geq 1\ \forall\ n \neq m$. Hence we get a contradiction. Thus every pseudo-Cauchy sequence of distinct points in (X,d) has a Cauchy subsequence. Hence by Proposition 1.2, $(\widehat{X}, d)$ is UC. □

Remark 5.3. (i) Theorem 5.4 still holds if we replace the metric space (Y,ρ) by $(\mathbb{R}, |\cdot|)$.
(ii) The equivalence of the conditions from (a) to (f) with (j) was proved in [Aggarwal and Kundu (2016)].
(iii) The conditions (a) and (g) were shown to be equivalent in [Beer and Garrido (2016)].

Recall that in Theorem 2.8 and Theorem 3.4, we have already seen some of the equivalent conditions which evolved while finding necessary and sufficient conditions on the domain space for pairwise coincidence of the Lipschitz-type functions given in [Beer and Garrido (2016, 2015)]. The next result was given in [Beer and Garrido (2015)] in the same spirit. It is immediately followed by Theorem 1.3 and Theorem 5.4 (g).

Corollary 5.1. *([Beer and Garrido (2015)]) Let (X,d) be a metric space. Then the following statements are equivalent:*

- (a) *(X,d) is a UC space.*
- (b) *Every locally Lipschitz function on (X,d) with values in an arbitrary metric space (Y,ρ) is Lipschitz in the small.*
- (c) *Every real-valued locally Lipschitz function on (X,d) is Lipschitz in the small.*

The following theorem characterizes those metric spaces on which every uniformly locally Lipschitz function with values in an arbitrary metric space is Lipschitz in the small. Interestingly, such metric spaces are precisely those whose completions are UC.

Theorem 5.5. *([Beer and Garrido (2015)]) Let (X,d) be a metric space. Then the following statements are equivalent:*

- (a) *The completion $(\widehat{X},d)$ of (X,d) is a UC space.*
- (b) *Each uniformly locally Lipschitz function on (X,d) with values in an arbitrary metric space (Y,ρ) is Lipschitz in the small.*
- (c) *Each real-valued uniformly locally Lipschitz function on (X,d) is Lipschitz in the small.*

Proof. $(a) \Rightarrow (b)$: Since every uniformly locally Lipschitz function between two metric spaces is Cauchy-Lipschitz, the implication follows from Theorem 5.4.

$(b) \Rightarrow (c)$: This is immediate.

$(c) \Rightarrow (a)$: Suppose (x_n) is a sequence in X with $\lim_{n\to\infty} I(x_n) = 0$, but has no Cauchy subsequence. Then $A = \{x_n : n \in \mathbb{N}\}$ is not totally bounded and hence by passing to a subsequence we can find $\delta > 0$ such that $d(x_k,x_n) \geq 2\delta \;\; \forall\, k \neq n$. Consequently, the function $f : (X,d) \to \mathbb{R}$ defined as:

$$f(x) = \left\{ \begin{array}{ll} k - \frac{3k}{\delta} d(x,x_k) & : x \in B(x_k, \frac{\delta}{3}) \text{ for some } k \in \mathbb{N} \\ 0 & : \text{otherwise} \end{array} \right\}$$

is uniformly locally Lipschitz as for all $x \in X$, $B(x, \frac{\delta}{3})$ intersects at most one of the balls $B(x_k, \frac{\delta}{3})$. Now we claim that f is not Lipschitz in the small. Since

$\lim_{n\to\infty} I(x_n) = 0$, for all n sufficiently large we can find $w_n \in X$ with $0 < d(x_n, w_n) < \frac{\delta}{3}$ and hence $|f(x_n) - f(w_n)| = \frac{3n}{\delta} d(x_n, w_n)$. Thus the claim is proved which gives a contradiction. □

Remark 5.4. One can find various other types of characterizations of UC spaces in the literature. For example, in terms of geometric functionals, sequences, functions, hyperspaces, etc [Atsuji (1958); Beer (1985, 1993, 2012); Beer *et al.* (2020, 2018); Chaves (1985); Kundu and Jain (2006); Monteiro and Peixoto (1951)].Various other characterizations of metric spaces having UC completion can be found in [Aggarwal and Kundu (2017b); Beer (1986); Beer and Garrido (2015); Beer *et al.* (2018); Jain and Kundu (2007)].

In Chapter 3, we have seen that every CC-regular function is Cauchy-subregular. Since every PC-regular function is CC-regular, every PC-regular function is also Cauchy-subregular. Let us see when the converse would be true.

Theorem 5.6. *[Gupta and Kundu (2022)] Let (X,d) be a metric space. Then the following assertions are equivalent.*

(*a*) *The completion $(\widehat{X},d)$ of (X,d) is UC space.*
(*b*) *Whenever (Y,ρ) is a metric space and $f : (X,d) \to (Y,\rho)$ is Cauchy-subregular, then f is PC-regular.*
(*c*) *Whenever (Y,ρ) is a metric space and $f : (X,d) \to (Y,\rho)$ is both continuous and Cauchy-subregular, then f is PC-regular.*
(*d*) *Whenever (Y,ρ) is a metric space and $f : (X,d) \to (Y,\rho)$ is both locally Lipschitz and Cauchy-subregular, then f is PC-regular.*
(*e*) *If $f : (X,d) \to \mathbb{R}$ is both locally Lipschitz and Cauchy-subregular, then f is PC-regular.*

Proof. In a manner similar to the proof of Theorem 3.8, we can prove the result. □

In [Beer *et al.* (2018)], it was proved that a metric space (X,d) is UC if and only if the reciprocal of every never zero real-valued uniformly continuous function is uniformly continuous. Our next result characterizes UC spaces in terms of stability under reciprocation of never zero PC-regular functions.

Theorem 5.7. *[Gupta and Kundu (2020)] Let (X,d) be a metric space. Then the following conditions are equivalent.*

(*a*) *(X,d) is a UC space.*

(*b*) *Whenever* $f:(X,d)\to\mathbb{R}$ *is a continuous PC-regular function such that* f *is never zero, then* $\frac{1}{f}$ *is also continuous and PC-regular.*

Proof. $(a)\Rightarrow(b)$: This follows from the fact that continuous function on a UC space is PC-regular.

$(b)\Rightarrow(a)$: If (X,d) is not UC, then there exists a pseudo-Cauchy sequence (x_n) of distinct points in (X,d) with no cluster point. Thus, $A=\{x_n : n\in\mathbb{N}\}$ is closed and discrete. Define

$$f:A\longrightarrow(0,2)$$

$$x_n\longmapsto\frac{1}{n}$$

Clearly, f is a continuous PC-regular function. By Theorem 1.1, f can be extended to a function $F:X\to(0,2)$, such that F is continuous. Since $(0,2)$ is totally bounded in $\mathbb{R}$, by Proposition 1.4, F is PC-regular. Cleary F is never zero but $\frac{1}{F}$ is not PC-regular, as (x_n) is pseudo-Cauchy but (n) is not pseudo-Cauchy in $\mathbb{R}$. We get a contradiction. □

We end this section by studying the condition under which a non-vanishing strongly uniformly continuous function on a set is stable under reciprocation. More precisely, if f is a never zero real-valued continuous function on (X,d) such that f is strongly uniformly continuous on a subset A of X, then under what conditions would $\frac{1}{f}$ be strongly uniformly continuous on A? To achieve our intended goal, let us recall the following definition.

Definition 5.2. [Beer and Levi (2009a)] A subset A of a metric space (X,d) is called a *UC set* if every sequence (a_n) in A with $\lim_{n\to\infty} I(a_n)=0$ has a cluster point in X.

We skip the routine proof of the following lemma.

Lemma 5.2. *[Beer and Levi (2009a)] Let* (X,d) *be a metric space and* $A\subseteq X$ *be a UC set. Then every real-valued continuous function on* X *is strongly uniformly continuous on* A.

Now we are ready to prove the desired result.

Theorem 5.8. *Let* (X,d) *be a metric space and* A *be a non-empty subset of* X. *Then the following statements are equivalent.*

(*a*) *A is a UC set.*

(b) *Whenever $f : (X,d) \to \mathbb{R}$ is a continuous and never zero function such that f is strongly uniformly continuous on A, then $\frac{1}{f}$ is also strongly uniformly continuous on A.*

Proof. $(a) \Rightarrow (b)$: This follows from Lemma 5.2.

$(b) \Rightarrow (a)$: Suppose A is not a UC set. Thus there exists a sequence (a_n) of distinct points in A such that $\lim_{n\to\infty} I(a_n) = 0$, but (a_n) has no cluster point in X. Without loss of generality, let us assume that $I(a_n) < \frac{1}{n}\ \forall n \in \mathbb{N}$.
<u>*Case 1:*</u> The sequence (a_n) is pseudo-Cauchy.

Let $B = \{a_n : n \in \mathbb{N}\}$. Let f be a real-valued function on B such that $f(a_n) = \frac{1}{n}\ \forall n \in \mathbb{N}$. Since f is a bounded uniformly continuous function on B, by Theorem 1.2, it can be extended to a uniformly continuous function $\hat{f}$ on the whole space X. Define the following function.

$$\begin{aligned} g : X &\longrightarrow \mathbb{R} \\ x &\longmapsto \hat{f}(x) + d(x,B) \end{aligned}$$

It is clear that g is uniformly continuous and is never zero. Thus g is strongly uniformly continuous on A, but $\frac{1}{g}$ is not uniformly continuous on A as (a_n) is pseudo-Cauchy and $g(a_n) = n\ \forall n \in \mathbb{N}$.
<u>*Case 2:*</u> The sequence (a_n) is not pseudo-Cauchy.

Thus there exist $\varepsilon > 0$ and $n'_o \in \mathbb{N}$ such that $d(a_n,a_m) > \varepsilon\ \forall n,m \geqslant n'_o$. Let $n_o = \max\{n'_o, \frac{4}{\varepsilon}+1\}$. Since $\forall n \in \mathbb{N}$, $I(a_n) < \frac{1}{n}$, $\exists x_n \in X$ such that $x_n \neq a_n$ and $d(a_n,x_n) < \frac{1}{n}$. Let $B = \{a_n, x_n : n \geq n_o\}$. Define a function f on B such that $f(x_n) = \frac{1}{n}$ and $f(a_n) = \frac{1}{n^2}\ \forall n \geqslant n_o$. Again, extend f to a uniformly continuous function $\hat{f}$ on X. Define the following function.

$$\begin{aligned} g : X &\longrightarrow \mathbb{R} \\ x &\longmapsto \hat{f}(x) + d(x,B) \end{aligned}$$

Thus g is strongly uniformly continuous on A as g is uniformly continuous. But $\frac{1}{g}$ is not strongly uniformly continuous on A as $d(a_n,x_n) < \frac{1}{n}$, but $|g(a_n) - g(x_n)| = n(n-1)\ \forall n \geqslant n_o$. □

The interested readers can also read [Ayala-Gómez *et al.* (2019); Beer (1988, 2020); Beer and Segura (2009); Doss (1947); Gutú and Jaramillo (2019); Keremedis (2018a,b)].

5.2 Cofinally Bourbaki-complete Metric Spaces

In Chapter 1, we have discussed that cofinal Bourbaki-completeness is associated with Bourbaki-completeness in the same way as cofinal completeness is associated with completeness. In this section, we focus on those functions which preserve cofinally Bourbaki-Cauchy sequences called CBC-regular functions. We find relation between CBC-regular and CC-regular functions and characterize the spaces on which the functions that are both CBC-regular and locally Lipschitz are bounded. In the process we obtain some characterizations of cofinally Bourbaki-complete spaces and finitely chainable spaces.

We start with some characterizations of cofinally Bourbaki-complete metric spaces using CBC-regular functions.

Theorem 5.9. *Let (X,d) be a metric space. Then the following statements are equivalent.*

- (a) *The metric space (X,d) is cofinally Bourbaki-complete.*
- (b) *Every d-cofinally Bourbaki-Cauchy sequence (that is, cofinally Bourbaki-Cauchy with respect to metric d) in X is σ-cofinally Bourbaki-Cauchy for all equivalent metrics σ on X.*
- (c) *Every continuous function on (X,d) with values in a metric space (Y,ρ) is CBC-regular.*
- (d) *Every locally Lipschitz function on (X,d) with values in a metric space (Y,ρ) is CBC-regular.*
- (e) *Every real-valued locally Lipschitz function on (X,d) is CBC-regular.*
- (f) *Every real-valued continuous function on (X,d) is CBC-regular.*

Proof. The implications $(a) \Rightarrow (b)$, $(c) \Rightarrow (d) \Rightarrow (e)$ are all immediate.

$(b) \Rightarrow (c)$: Let $f : (X,d) \to (Y,\rho)$ be a continuous function. Now, define a metric σ on X as follows:

$$\sigma(a,b) = d(a,b) + \rho(f(a), f(b)) \text{ for } a,\ b \text{ in } X$$

Then σ is equivalent to d. Let (x_n) be a d-cofinally Bourbaki-Cauchy sequence in X, then it is σ-cofinally Bourbaki-Cauchy. Hence by the definition of σ, $(f(x_n))$ is cofinally Bourbaki-Cauchy in (Y,ρ).

$(e) \Rightarrow (f)$: Let $f : (X,d) \to \mathbb{R}$ be a continuous function and let (x_n) be a cofinally Bourbaki-Cauchy sequence in (X,d). Let $\varepsilon > 0$. By Theorem 1.4, there exists a locally Lipschitz function $g : (X,d) \to \mathbb{R}$ with $\sup_{x \in X} |f(x) - g(x)| < \varepsilon$. By (e), g is CBC-regular and hence $(g(x_n))$ is a cofinally Bourbaki-Cauchy sequence in $\mathbb{R}$. Consequently, there exist $m \in \mathbb{N}$ and an infinite subset N of $\mathbb{N}$ such that $g(x_k)$ and $g(x_n)$ can be joined by an ε-chain of length m for all $k,\ n \in N$. Thus $f(x_k)$ and

$f(x_n)$ can be joined by an ε-chain of length $m+2$ for all k, $n \in N$. Since ε was arbitrary, $(f(x_n))$ is cofinally Bourbaki-Cauchy in $\mathbb{R}$ and hence f is CBC-regular.

$(f) \Rightarrow (a)$: Suppose that (x_n) is a cofinally Bourbaki-Cauchy sequence in (X,d). If (x_n) has a constant subsequence then we are done, or else it has a cofinally Bourbaki-Cauchy subsequence of distinct points. Thus it suffices to prove that every cofinally Bourbaki-Cauchy sequence in (X,d) with distinct terms clusters. So let (x_n) be a sequence of distinct points which does not cluster. Then the set $A = \{x_n : n \in \mathbb{N}\}$ is closed and discrete and hence the function $f : A \to \mathbb{R}$: $f(x_n) = 2^n$ is continuous on (A,d). By Tietze's extension theorem, f can be extended to a real-valued continuous function f' on (X,d) which is not CBC-regular. We get a contradiction. Thus, (X,d) is cofinally Bourbaki-complete. □

Remark 5.5. The equivalence of the conditions (c) and (f) with (a) was proved in [Aggarwal and Kundu (2017a)].

Let us note some characterizations of metric spaces having cofinally Bourbaki-complete completion.

Theorem 5.10. *Let (X,d) be a metric space. Then the following statements are equivalent.*

- (a) *The completion $(\widehat{X},d)$ of (X,d) is cofinally Bourbaki-complete.*
- (b) *Every d-cofinally Bourbaki-Cauchy sequence in X is σ-cofinally Bourbaki-Cauchy for all Cauchy equivalent metrics σ on X.*
- (c) *Every Cauchy-regular function on (X,d) with values in a metric space (Y,ρ) is CBC-regular.*
- (d) *Every Cauchy-Lipschitz function on (X,d) with values in a metric space (Y,ρ) is CBC-regular.*
- (e) *Every real-valued Cauchy-Lipschitz function on (X,d) is CBC-regular.*
- (f) *Every real-valued Cauchy-regular function on (X,d) is CBC-regular.*
- (g) *Every cofinally Bourbaki-Cauchy sequence in (X,d) has a Cauchy subsequence.*

Proof. The implications $(a) \Rightarrow (b)$ and $(c) \Rightarrow (d) \Rightarrow (e)$ are immediate.

$(b) \Rightarrow (c)$: Let $f : (X,d) \to (Y,\rho)$ be a Cauchy-regular function. Now, define a metric σ on X as follows:

$$\sigma(a,b) = d(a,b) + \rho(f(a), f(b)) \text{ for } a,\ b \text{ in } X$$

Then σ is Cauchy equivalent to d by Proposition 3.1. Let (x_n) be a d-cofinally Bourbaki-Cauchy sequence in X, then it is σ-cofinally Bourbaki-Cauchy. Hence by the definition of σ, $(f(x_n))$ is cofinally Bourbaki-Cauchy in (Y,ρ).

$(e) \Rightarrow (f)$: It follows from Theorem 1.5.

$(f) \Rightarrow (g)$: Let (x_n) be a cofinally Bourbaki-Cauchy sequence in (X,d) with no Cauchy subsequence. We can assume that (x_n) has distinct terms. Let $A = \{x_n : n \in \mathbb{N}\}$. Consider the function $f : (A,d) \to \mathbb{R}$: $f(x_n) = 2^n$. Then f is a Cauchy-regular function on (A,d). Hence by Corollary 1.1, f can be extended to a real-valued Cauchy-regular function f' on (X,d), which is not CBC-regular. We arrive at a contradiction.

$(g) \Rightarrow (a)$: We skip the routine proof of this implication. □

Remark 5.6. The equivalence of the conditions (c), (f), (g) with (a) was proved in [Aggarwal and Kundu (2017a)].

By the previous theorem, we can conclude that a real-valued Cauchy-regular function need not be CBC-regular. Using Proposition 1.5 and the previous result, the next corollary can be easily proved.

Corollary 5.2. *The following statements are equivalent for a metric space (X,d):*

- (a) *The metric space (X,d) is totally bounded.*
- (b) *The metric space (X,d) is finitely chainable and its completion is cofinally Bourbaki-complete.*

Recall that in Theorem 2.11, we have seen that a function between two metric spaces is CC-regular if and only if it is almost bounded. Here we attempt to give a similar type of characterization for CBC-regular functions.

Theorem 5.11. *[Gupta and Kundu (2020)] Let f be a function from a metric space (X,d) to another metric space (Y,ρ). Then the following conditions are equivalent.*

- (a) *f is CBC-regular.*
- (b) *For every $\varepsilon > 0$, there exists $\delta > 0$ such that for all $x \in X$ and $m \in \mathbb{N}$, there exist $n \in \mathbb{N}$ and $\{y_1, y_2, \ldots, y_k\} \subseteq Y$ such that $f\Big(B^m(x,\delta)\Big) \subseteq \bigcup_{i=1}^{k} B^n(y_i,\varepsilon)$.*

Proof. $(a) \Rightarrow (b)$: Suppose $\exists \varepsilon > 0$ such that $\forall \delta > 0$, $\exists x_\delta$ and m_x such that $f\Big(B^{m_x}(x_\delta,\delta)\Big)$ cannot be bounded by any finite union of $n^{th}\varepsilon$-enlarged open balls. For each $n \in \mathbb{N}$, let $\delta_n = \frac{1}{n}$, thus there exists x_n and m_{x_n} (say m_n) such that $f\Big(B^{m_n}(x_n,\frac{1}{n})\Big)$ cannot be contained in any finite union of the type $\bigcup_{i=1}^{k} B^l(y_i,\varepsilon)$ for all k, $l \in \mathbb{N}$. Let $a_1^n \in f\Big(B^{m_n}(x_n,\frac{1}{n})\Big)$. Since $f\Big(B^{m_n}(x_n,\frac{1}{n})\Big) \subsetneq B(a_1^n,\varepsilon)$, there exists $a_2^n \in f\left(B^{m_n}(x_n, \frac{1}{n})\right)$ such that $a_2^n \notin B(a_1^n, \varepsilon)$. Similarly, there exists $a_3^n \in f\left(B^{m_n}(x_n, \frac{1}{n})\right)$ such that $a_3^n \notin B^2(a_1^n, \varepsilon) \cup B^2(a_2^n, \varepsilon)$. Thus we get a sequence $(a_m^n)_{m \in \mathbb{N}} \subseteq f\left(B^{m_n}(x_n, \frac{1}{n})\right)$ such that a_j^n cannot be bound with a_i^n by an ε-chain of length $j-1$ for $0 < i < j$. Let $A_n = \{a_m^n : m \in \mathbb{N}\}$. Now we will construct a sequence in Y. Let $F_1 = \{a_{\alpha_1}^1\}$ where $\alpha_1 \geqslant 1$. Let $a_{\alpha_2}^2 \in A_2$ ($\alpha_2 \geqslant 2$) such that $a_{\alpha_2}^2$ cannot be joined with $a_{\alpha_1}^1$ by an ε-chain of length 1. It is possible because if there does not exist such $a_{\alpha_2}^2$, then all the elements $(a_m^n)_{m \geqslant max\{\alpha_1, 2\}}$ can be joined by an ε-chain of length 2, but it is a contradiction. Similarly, choose $a_{\alpha_3}^1 \in A_1$ ($\alpha_3 \geqslant 3$) such that $a_{\alpha_3}^1$ cannot be joined by an ε-chain of length 2 with the elements $a_{\alpha_1}^1$ and $a_{\alpha_2}^2$. Let $F_2 = \{a_{\alpha_1}^1, a_{\alpha_2}^2, a_{\alpha_3}^1\}$. Now suppose a finite subset $F_n = \left\{a_{\alpha_1}^1, a_{\alpha_2}^2, a_{\alpha_3}^1, \ldots, a_{\alpha_{\frac{n(n+1)}{2}}}^1\right\}$ of $\bigcup_{i=1}^{n} A_i$ is chosen such that $a_{\alpha_j}^l$ cannot be bound with $a_{\alpha_i}^m$ by an ε-chain of length $j-1$ for all $0 < i < j$ ($1 \leqslant l, m \leqslant n$) and $|F_n \cap A_i| = n - i + 1$ for all $1 \leq i \leq n$. Now we construct F_{n+1} as follows. First we claim that there exists $a_{\alpha_{\frac{n(n+1)}{2}+1}}^{n+1} \in A_{n+1}$ such that it cannot be joined with any member of F_n by an ε-chain of length $\frac{n(n+1)}{2}$. Let $k \geqslant \max\{\alpha_{\frac{n(n+1)}{2}}, \frac{n(n+1)}{2} + 1\}$. Consider $\{a_m : m \geq 2k\}$. Since $\{a_m : m \geq 2k\}$ is infinite and F_n is finite, there exists $y, y' \in \{a_m : m \geqslant 2k\} \backslash F_n$, $y \neq y'$, such that y and z can be joined by an ε-chain of length $\frac{n(n+1)}{2}$ and y' and z can also be joined by an ε-chain of length $\frac{n(n+1)}{2}$. Thus y and y' can be joined by an ε-chain of length $n(n+1)$. We get a contradiction. Repeating this process for $i = n, n-1, \ldots, 1$ and for the set $F_n \cup \left\{a_{\alpha_{\frac{n(n+1)}{2}+1}}^{n+1}, a_{\alpha_{\frac{n(n+1)}{2}+2}}^{n}, \ldots, a_{\alpha_{\frac{n(n+1)}{2}+(n-i+2)}}^{i}\right\}$ together with the infinite set E_i, we get the desired set F_{n+1}. Now, if we choose a sequence in the order we picked the elements, that is, $a_{\alpha_1}^1$, $a_{\alpha_2}^2$, $a_{\alpha_3}^1$, $a_{\alpha_4}^3, \ldots$, then this sequence is not cofinally Bourbaki-Cauchy, but its preimage is cofinally Bourbaki-Cauchy. We get a contradiction.

$(b) \Rightarrow (a)$: Let (x_n) be a cofinally Bourbaki-Cauchy sequence. We claim that $(f(x_n))$ is cofinally Bourbaki-Cauchy. Let $\varepsilon > 0$. Thus $\exists \delta > 0$ such that $\forall x \in X$ and $m \in \mathbb{N}$, $\exists n \in \mathbb{N}$ and $\{y_1, y_2, \ldots, y_k\} \subseteq Y$ such that $f\left(B^m(x, \delta)\right) \subseteq \bigcup_{i=1}^{k} B^n(y_i, \varepsilon)$. Since (x_n) is cofinally Bourbaki-Cauchy, there exist $m \in \mathbb{N}$ and an infinite subset $\mathbb{N}_\delta$ of $\mathbb{N}$ such that the points x_j and x_n can be joined by an δ-chain of length m for every j, $n \in \mathbb{N}_\delta$. Let $t \in \mathbb{N}_\delta$ and consider $B^m(x_t, \delta)$. Therefore, $\exists n \in \mathbb{N}$ and $\{y_1, y_2, \ldots, y_k\} \subseteq Y$ such that $f\left(B^m(x_t, \delta)\right) \subseteq \bigcup_{i=1}^{k} B^n(y_i, \varepsilon)$. At least one of these

$B^n(y_i, \varepsilon)$ will contain elements of the type $f(x_k)$ for infinite values of k. Hence $(f(x_n))$ is cofinally Bourbaki-Cauchy. □

Theorem 5.12. *[Gupta and Kundu (2020)] Let (X,d) be a metric space. Then the following assertions are equivalent.*

(a) *(X,d) is cofinally Bourbaki-complete.*
(b) *Whenever $f:(X,d) \to \mathbb{R}$ is a continuous CBC-regular function such that f is never zero, then $\frac{1}{f}$ is also continuous and CBC-regular.*

Proof. $(a) \Rightarrow (b)$: This follows from the fact that every real-valued continuous function on a cofinally Bourbaki-complete metric space is CBC-regular.

$(b) \Rightarrow (a)$: If (X,d) is not cofinally Bourbaki-complete, then there exists a cofinally Bourabki-Cauchy sequence (x_n) of distinct points in (X,d) with no cluster point. Thus, $A = \{x_n : n \in \mathbb{N}\}$ is closed and discrete. Define a function f from A to $\mathbb{R}$ such that $f(x_n) = \frac{1}{n}$ $\forall n \in \mathbb{N}$. Clearly, f is continuous and CBC-regular. By Theorem 1.1, there exists a continuous function $F : X \to (0,2)$ which extends f. Now $(0,2)$ is finitely chainable in $\mathbb{R}$ and thus by Proposition 1.5, F is CBC-regular. Now, F is never zero but $\frac{1}{F}$ is not CBC-regular, as (x_n) is cofinally Bourbaki-Cauchy but (n) is not cofinally Bourbaki-Cauchy in $\mathbb{R}$. □

We refer the interested reader to [Aggarwal and Kundu (2017b); Beer and Garrido (2020); Garrido and Meroño (2016); Garrido and Meroño (2014)] for more characterizations of cofinally Bourbaki-complete and Bourbaki-complete metric spaces. Recall that in Chapter 1, we have seen through examples that a CC-regular function need not be CBC-regular and a CBC-regular function need not be CC-regular. Similar behaviour was observed between BC-regular and CBC-regular functions. Next, we establish some relations between the aforesaid functions. In the process, we find some equivalent conditions under which cofinally complete metric spaces and Bourbaki-complete metric spaces are cofinally Bourbaki-complete.

Theorem 5.13. *[Gupta and Kundu (2020)] Let (X,d) be a metric space. Then the following statements are equivalent.*

(a) *(X,d) is cofinally Bourbaki-complete.*
(b) *(X,d) is cofinally complete and every CC-regular function from (X,d) to any other metric space (Y,ρ) is CBC-regular.*
(c) *(X,d) is cofinally complete and every real-valued CC-regular function on (X,d) is CBC-regular.*

(*d*) *(X,d) is cofinally complete and every cofinally Bourbaki-Cauchy sequence in (X,d) is cofinally Cauchy.*

Proof. The implications $(b) \Rightarrow (c)$ and $(d) \Rightarrow (a)$ are immediate.

$(a) \Rightarrow (b)$: Let $f : (X,d) \to (Y,\rho)$ be a CC-regular function. Let (x_n) be a cofinally Bourbaki-Cauchy sequence. Since (X,d) is cofinally Bourbaki-complete, there exists a convergent subsequence (x_{n_k}) of (x_n). This implies (x_{n_k}) is cofinally Cauchy and so is $(f(x_{n_k}))$. Thus, $(f(x_n))$ is cofinally Bourbaki-Cauchy, which implies f is CBC-regular.

$(c) \Rightarrow (d)$: Suppose there exists a cofinally Bourbaki-Cauchy sequence (x_n) of distinct points in (X,d) such that it is not cofinally Cauchy. Now, define a function $f : X \to \mathbb{R}$ as:

$$f(x) = \begin{cases} n : x = x_n \text{ for some } n \in \mathbb{N} \\ 0 : \text{otherwise} \end{cases}$$

Clearly, f is CC-regular but not CBC-regular, a contradiction. □

Theorem 5.14. *[Gupta and Kundu (2020)] Let (X,d) be a metric space. Then the following conditions are equivalent.*

(*a*) *$(\widehat{X},d)$ is cofinally Bourbaki-complete.*
(*b*) *$(\widehat{X},d)$ is cofinally complete and every CC-regular function from (X,d) to any other metric space (Y,ρ) is CBC-regular.*
(*c*) *$(\widehat{X},d)$ is cofinally complete and every real-valued CC-regular function on (X,d) is CBC-regular.*
(*d*) *$(\widehat{X},d)$ is cofinally complete and every cofinally Bourbaki-Cauchy sequence in (X,d) is cofinally Cauchy.*

Proof. $(a) \Rightarrow (b)$: Since $(\widehat{X},d)$ is cofinally Bourbaki-complete, every cofinally Bourbaki-Cauchy sequence in it has a Cauchy subsequence and hence is cofinally Cauchy.

$(b) \Rightarrow (c)$: This is immediate.

$(c) \Rightarrow (d)$: This can be proved in a manner similar to the proof of $(c) \Rightarrow (d)$ in Theorem 5.13.

$(d) \Rightarrow (a)$: Suppose $(\widehat{X},d)$ is not cofinally Bourbaki-complete. Thus there exists a cofinally Bourbaki-Cauchy sequence $(\hat{x_n})$ in $\widehat{X}$ which has no Cauchy subsequence. Since X is dense in $\widehat{X}$, for each $n \in \mathbb{N}$, there exists $x_n \in X$ such that $d(x,\hat{x}) < \frac{1}{n}$. Thus (x_n) is a cofinally Bourbaki-Cauchy sequence in (X,d) without any Cauchy subsequence. By hypothesis, we get a cofinally Cauchy sequence (x_n) in $(\widehat{X},d)$ without any cluster point, which is a contradiction to the fact that $(\widehat{X},d)$ is cofinally complete. Hence $(\widehat{X},d)$ is cofinally Bourbaki-complete. □

Remark 5.7. Note that if a metric space (X,d) or its completion $(\widehat{X},d)$ is cofinally Bourbaki-complete, then every cofinally Bourbaki-Cauchy sequence in (X,d) is cofinally Cauchy. But the converse need not hold. For example, as a metric subspace of the Hilbert space ℓ_2, consider $X = \{e_n + \frac{1}{n}e_k : n,\ k \in \mathbb{N}\}$. Then (X,d) is complete and every cofinally Bourbaki-Cauchy sequence in (X,d) is cofinally Cauchy (the metric d is induced by the ℓ_2-norm), but (X,d) is not cofinally Bourbaki-complete. Thus the conditions that (X,d) is cofinally complete and $(\widehat{X},d)$ is cofinally complete in Theorem 5.13 and Theorem 5.14 respectively, cannot be dropped.

The proof of the following result is analogous to that of Theorem 5.13.

Theorem 5.15. *[Gupta and Kundu (2020)] Let (X,d) be a metric space. Then the following assertions are equivalent.*

(a) (X,d) is cofinally Bourbaki-complete.
(b) (X,d) is Bourbaki-complete and every BC-regular function from (X,d) to any other metric space (Y,ρ) is CBC-regular.
(c) (X,d) is Bourbaki-complete and every real-valued BC-regular function on (X,d) is CBC-regular.
(d) (X,d) is Bourbaki-complete and every cofinally Bourbaki-Cauchy sequence in (X,d) has a Bourbaki-Cauchy subsequence.

In a manner similar to the proof of Theorem 5.14, we can prove the following result.

Theorem 5.16. *[Gupta and Kundu (2020)] Let (X,d) be a metric space. Then the following statements are equivalent.*

(a) $(\widehat{X},d)$ is cofinally Bourbaki-complete.
(b) $(\widehat{X},d)$ is Bourbaki-complete and every BC-regular function from (X,d) to any other metric space (Y,ρ) is CBC-regular.
(c) $(\widehat{X},d)$ is Bourbaki-complete and every real-valued BC-regular function on (X,d) is CBC-regular.
(d) $(\widehat{X},d)$ is Bourbaki-complete and every cofinally Bourbaki-Cauchy sequence in (X,d) has a Bourbaki-Cauchy subsequence.

Remark 5.8. Note that if a metric space (X,d) or its completion $(\widehat{X},d)$ is cofinally Bourbaki-complete, then every cofinally Bourbaki-Cauchy sequence in (X,d) has a Bourbaki-Cauchy subsequence. But the converse need not hold. For example, consider the real Hilbert space ℓ_2. Let $X \subseteq \ell_2$ be the closed ball around 0 of radius 2. Then the metric space (X,d) is finitely chainable. Thus by Proposition 1.5, every cofinally Bourbaki-Cauchy sequence in it has a Bourbaki-Cauchy subsequence. (X,d) is complete but not cofinally Bourbaki-complete because if we enumerate the elements of $A = \{e_n + \frac{1}{n}e_k : n,\ k \in \mathbb{N}\} \subset X$, we will get a cofinally Bourbaki-Cauchy sequence which does not have any cluster point in X. Thus the conditions that (X,d) is Bourbaki-complete and $(\widehat{X},d)$ is Bourbaki-complete in Theorem 5.15 and Theorem 5.16 respectively, cannot be dropped.

In 1958, Atsuji proved that every real-valued uniformly continuous function on a metric space (X,d) is bounded if and only if (X,d) is finitely chainable [Atsuji (1958)]. It is known that the finite chainability of (X,d) is also characterized by the boundedness of Lipschitz in the small functions on the space. This is implicit in Atsuji's proof and explicitly proved in [Beer and Garrido (2014)]. In [Beer and Garrido (2014)], it was also proved that the boundedness of other Lipschitz-type functions characterize total boundedness of the domain space. In our next result, we study the boundedness of some combinations of Lipschitz-type functions with CBC-regular functions. More precisely, the next result shows that the metric spaces on which every real-valued continuous CBC-regular function is bounded are exactly the finitely chainable metric spaces.

Theorem 5.17. *[Gupta and Kundu (2020)] Let (X,d) be a metric space. Then the following statements are equivalent.*

- (a) *(X,d) is finitely chainable.*
- (b) *Whenever (Y,ρ) is a metric space and $f : (X,d) \to (Y,\rho)$ is both continuous and CBC-regular, then f is bounded.*
- (c) *Whenever (Y,ρ) is a metric space and $f : (X,d) \to (Y,\rho)$ is both locally Lipschitz and CBC-regular, then f is bounded.*
- (d) *Whenever (Y,ρ) is a metric space and $f : (X,d) \to (Y,\rho)$ is both Cauchy-Lipschitz and CBC-regular, then f is bounded.*
- (e) *Whenever (Y,ρ) is a metric space and $f : (X,d) \to (Y,\rho)$ is both uniformly locally Lipschitz and CBC-regular, then f is bounded.*
- (f) *If $f : (X,d) \to \mathbb{R}$ is both uniformly locally Lipschitz and CBC-regular, then f is bounded.*

Proof. The implications $(b) \Rightarrow (c) \Rightarrow (d) \Rightarrow (e) \Rightarrow (f)$ are all immediate.

$(a) \Rightarrow (b)$: Let f be a continuous CBC-regular function from (X,d) to (Y,ρ). By Proposition 1.5, $f(X)$ is finitely chainable in (Y,ρ).

$(f) \Rightarrow (a)$: Suppose (X,d) is not finitely chainable. Therefore, there exists $\varepsilon > 0$ such that for any finite subset $\{x_1, x_2, \dots, x_n\}$ of X and for every $m \in \mathbb{N}$, there exists $x \in X$ such that x cannot be joined with any x_i by an ε-chain of length m. Consider $a_1 \in X$. Then for $m = 1$, there exists $a_2 \in X$ such that a_2 cannot be bound with a_1 by an ε-chain of length 1. By induction there exists $\{a_n : n \in \mathbb{N}\}$ where a_j cannot be bound with a_i by an ε-chain of length $j-1$, for $0 < i < j$. Define a function $f : (X,d) \to \mathbb{R}$ as follows:

$$f(x) = \begin{cases} n - \frac{4n}{\varepsilon} d(x, a_n) : x \in B\left(a_n, \frac{\varepsilon}{4}\right) \text{ for some } n \in \mathbb{N} \\ 0 \qquad\qquad\qquad : \text{otherwise} \end{cases}$$

Then, by giving a similar explanation as given in Theorem 2.18, the function f is uniformly locally Lipschitz. To see f is CBC-regular, let (z_n) be a cofinally Bourbaki-Cauchy sequence in (X,d). If there exists an infinite subset N' of $\mathbb{N}$ such that $z_k \notin \bigcup_{i \in \mathbb{N}} B\left(a_i, \frac{\varepsilon}{4}\right)\ \forall k \in N'$, then $(f(z_n))$ is a cofinally Bourbaki-Cauchy sequence as $f(z_k) = 0\ \forall k \in N'$. If that is not the case, then there exists an infinite subset N' of $\mathbb{N}$ such that $\forall k \in N'$, $z_k \in B(a_l, \frac{\varepsilon}{4})$ for some $l \in \mathbb{N}$, because otherwise for $\frac{\varepsilon}{4} > 0$, there would not be any infinite subset $N_{\frac{\varepsilon}{4}}$ of $\mathbb{N}$ and $m \in \mathbb{N}$ such that the points z_j and z_n can be joined by an $\frac{\varepsilon}{4}$-chain of length m for every $j, n \in \mathbb{N}_{\frac{\varepsilon}{4}}$, which contradicts the fact that (z_n) is cofinally Bourbaki-Cauchy. Thus for some $t \geq l$, $f(z_k) \in [0,t]\ \forall k \in N'$. Enumerate the elemets of N' in increasing order and let $N' = \{k_1, k_2, k_3, \ldots\}$. Since $[0,t]$ is finitely chainable in $\mathbb{R}$, $(f(z_{k_n}))$ is cofinally Bourbaki-Cauchy. Thus $(f(z_n))$ is cofinally Bourbaki-Cauchy. Hence f is CBC-regular and uniformly locally Lipschitz but unbounded. We get a contradiction. □

For additional reading, one may refer to [Beer (1981); Bouziad and Sukhacheva (2019); Garrido and Meroño (2012, 2013)].

Exercises

Exercise 5.1

Prove Proposition 5.1 and Proposition 5.2.

Exercise 5.2

Give examples of:

(a) a cofinally Bourbaki-Cauchy sequence which is not cofinally Cauchy.
(b) a cofinally Bourbaki-Cauchy sequence which has no Bourbaki-Cauchy subsequence.
(c) a metric space (X,d) which is not totally bounded but its completion is cofinally Bourbaki-complete.

Exercise 5.3

[Beer *et al.* (2020)] Let (X,d) be a metric space. Then prove the equivalence of the following assertions.

(a) (X,d) is a UC space.
(b) Whenever $f:(X,d)\to\mathbb{R}$ is Lipschitz in the small such that f is never zero, then $\frac{1}{f}$ is also Lipschitz in the small.
(c) Whenever $f:(X,d)\to\mathbb{R}$ is Lipschitz such that f is never zero, then $\frac{1}{f}$ is Lipschitz in the small.
(d) [Beer *et al.* (2018)] Whenever $f:(X,d)\to\mathbb{R}$ is uniformly continuous such that f is never zero, then $\frac{1}{f}$ is also uniformly continuous.

Exercise 5.4

[Garrido and Meroño (2014)] Show that every uniformly locally compact metric space is cofinally Bourbaki-complete.

Exercise 5.5

[Garrido and Meroño (2014)] Let (X,d) be a metric space. Prove the following are equivalent.

(a) X is uniformly locally compact.
(b) X is locally totally bounded and cofinally complete.
(c) X is locally Bourbaki-bounded and cofinally Bourbaki-complete.

Exercise 5.6

[Aggarwal and Kundu (2017b)] Let (X,d) be a metric space. A sequence (x_n) is said to be *pseudo-Bourbaki-Cauchy* in X if for every $\varepsilon>0$, there exists $m\in\mathbb{N}$ such that $\forall\, n\in\mathbb{N}$, $\exists\, j,\, k>n,\, j\neq k$ such that the points x_j and x_k can be joined by an ε-chain of length m. Prove that the following statements are equivalent.

(a) The metric space (X,d) is a UC space.
(b) Every d-pseudo-Bourbaki-Cauchy sequence in X is σ-pseudo-Bourbaki-Cauchy for all equivalent metrics σ on X.
(c) Every pseudo-Bourbaki-Cauchy sequence in (X,d) with distinct terms clusters.
(d) If (x_n) and (y_n) are cofinally Bourbaki-asymptotic sequences in X such that $x_n\neq y_n$ for each n, then the sequences (x_n) and (y_n) cluster in X.
(e) If (x_n) and (y_n) are Bourbaki-asymptotic sequences in X such that $x_n\neq y_n$ for each n, then the sequences (x_n) and (y_n) cluster in X.

(f) Each continuous function on (X,d) with values in a metric space (Y,ρ) preserves pairs of cofinally Bourbaki-asymptotic sequences.
(g) Each continuous function on (X,d) with values in a metric space (Y,ρ) preserves pairs of Bourbaki-asymptotic sequences.
(h) Each locally Lipschitz function on (X,d) with values in a metric space (Y,ρ) preserves pairs of Bourbaki-asymptotic sequences.
(i) Each locally Lipschitz function on (X,d) with values in a metric space (Y,ρ) preserves pairs of cofinally Bourbaki-asymptotic sequences.

Exercise 5.7

[Beer and Garrido (2014, 2015); Kundu *et al.* (2017)] Let (X,d) be a metric space. Then prove that the following assertions are equivalent:

(a) (X,d) is finitely chainable.
(b) Every countable subset of X is finitely chainable in (X,d).
(c) Whenever (Y,ρ) is a metric space and $f:(X,d)\to(Y,\rho)$ is BC-regular, then f is bounded.
(d) Whenever (Y,ρ) is a metric space and $f:(X,d)\to(Y,\rho)$ is CBC-regular, then f is bounded.
(e) Whenever (Y,ρ) is a metric space and $f:(X,d)\to(Y,\rho)$ is Lipschitz in the small, then f is bounded.
(f) (X,d) is bounded and whenever (Y,ρ) is a metric space and $f:(X,d)\to(Y,\rho)$ is Lipschitz in the small, then f is Lipschitz.
(g) Whenever (Y,ρ) is a metric space and $f:(X,d)\to(Y,\rho)$ is Lipschitz in the small, then f is both Lipschitz and bounded.

Exercise 5.8

[Kundu *et al.* (2017)] Verify that the following statements are equivalent for a metric space (X,d):

(a) (X,d) is totally bounded.
(b) Whenever (Y,ρ) is a metric space and $f:(X,d)\to(Y,\rho)$ is PC-regular, then f is bounded.
(c) (X,d) is finitely chainable and its completion is Bourbaki-complete.
(d) (X,d) is finitely chainable and its completion is a UC space.
(e) (X,d) is finitely chainable and every CC-regular function on (X,d) with values in any arbitrary chainable metric space (Y,ρ) is BC-regular.
(f) (X,d) is finitely chainable and every PC-regular function on (X,d) with values in any arbitrary chainable metric space (Y,ρ) is BC-regular.

(g) (X,d) is finitely chainable and every CC-regular function on (X,d) with values in any arbitrary metric space (Y,ρ) is CBC-regular.
(h) (X,d) is finitely chainable and every PC-regular function on (X,d) with values in any arbitrary metric space (Y,ρ) is CBC-regular.
(i) (X,d) is finitely chainable and every uniformly locally Lipschitz function on (X,d) with values in any arbitrary metric space (Y,ρ) is Lipschitz in the small.
(j) (X,d) is finitely chainable and every uniformly locally Lipschitz function on (X,d) with values in any arbitrary metric space (Y,ρ) is uniformly continuous.
(k) (X,d) is finitely chainable and every uniformly locally Lipschitz function on (X,d) with values in any arbitrary metric space (Y,ρ) is BC-regular.
(l) (X,d) is finitely chainable and every uniformly locally Lipschitz function on (X,d) with values in any arbitrary metric space (Y,ρ) is CBC-regular.

Exercise 5.9

[Cabello Sánchez (2017)] Let (X,d) be a metric space. Then the set of real-valued uniformly continuous functions on (X,d) is closed under pointwise product if and only if every subset A of X is either finitely chainable or contains a subset B such that $\inf\{I(b) : b \in B\} > 0$.

List of Symbols

Symbol	Meaning
$\forall$	for all
$\exists$	there exists
$\in$	belongs to
$\notin$	does not belong to
$\subseteq$	subset or equal
$\cup, \cap$	union, intersection
$X \setminus E$ or E^c	the complement of E in X
$\emptyset$	empty set
$\mathbb{N}$	the set of natural numbers
$\mathbb{Q}$	the set of rational numbers
$\mathbb{R}$	the real line
l_2	the Hilbert space of square summable real sequences
(e_n)	the standard orthonormal basis in l_2
$\square$	end of a proof
$\lvert \cdot \rvert$	the usual distance metric on $\mathbb{R}$
$\lVert \cdot \rVert$	norm
$(x_n)_{n\in\mathbb{N}}$	a sequence in a non-empty set, occasionally it may be denoted by (x_n)
$f\vert_A$	the restriction of f to A where $f : X \to Y$ is a function, $\emptyset \neq A \subseteq X$

For the following notations, (X,d) is a metric space, A is a non-empty subset of X, $x \in X$, and $\varepsilon > 0$

$\overline{A}$ or $cl_X A$	the closure of A in X
$int\,A$ or A°	the interior of A in X
X'	the set of accumulation points in X
$(\widehat{X},d)$	the completion of a metric space (X,d)
$B_d(x,\varepsilon)$ or $B(x,\varepsilon)$ or $B^1(x,\varepsilon)$	the open ball in (X,d), centered at $x \in X$ with radius ε
$C_d(x,\varepsilon)$ or $C(x,\varepsilon)$	the closed ball in (X,d), centered at $x \in X$ with radius ε
$d(x,A)$	distance between x and A which is $\inf\{d(x,a) : a \in A\}$
A^ε or $B(A,\varepsilon)$	the ε-enlargement of A, that is, the set $\{x \in X : d(x,A) < \varepsilon\}$
for every $n \geq 2$, $B^n(x,\varepsilon)$	the ε-enlargement of the set $B^{n-1}(x,\varepsilon)$

Bibliography

Aggarwal, M. and Kundu, S. (2016). More about the cofinally complete spaces and the Atsuji spaces, *Houston J. Math.* **42**, 4, pp. 1373–1395.

Aggarwal, M. and Kundu, S. (2017a). Boundedness of the relatives of uniformly continuous functions, *Topology Proc.* **49**, pp. 105–119.

Aggarwal, M. and Kundu, S. (2017b). More on variants of complete metric spaces, *Acta Math. Hungar.* **151**, 2, pp. 391–408.

Atsuji, M. (1958). Uniform continuity of continuous functions of metric spaces, *Pacific J. Math.* **8**, pp. 11–16.

Ayala-Gómez, R., Bernal-González, L., Calderón-Moreno, M. C., and Vilches-Alarcón, J. A. (2019). Structural aspects of the non-uniformly continuous functions and the unbounded functions within $C(X)$, *J. Math. Anal. Appl.* **472**, 1, pp. 372–385.

Beer, G. (1981). Which connected metric spaces are compact? *Proc. Amer. Math. Soc.* **83**, 4, pp. 807–811.

Beer, G. (1985). Metric spaces on which continuous functions are uniformly continuous and Hausdorff distance, *Proc. Amer. Math. Soc.* **95**, 4, pp. 653–658.

Beer, G. (1986). More about metric spaces on which continuous functions are uniformly continuous, *Bull. Austral. Math. Soc.* **33**, 3, pp. 397–406.

Beer, G. (1988). UC spaces revisited, *Amer. Math. Monthly* **95**, 8, pp. 737–739.

Beer, G. (1993). *Topologies on closed and closed convex sets, Mathematics and its Applications*, Vol. 268 (Kluwer Academic Publishers Group, Dordrecht).

Beer, G. (2008). Between compactness and completeness, *Topology Appl.* **155**, 6, pp. 503–514.

Beer, G. (2012). Between the cofinally complete spaces and the UC spaces, *Houston J. Math.* **38**, 3, pp. 999–1015.

Beer, G. (2020). On boundedly compact metrics and UC metrics, *Bull. Belg. Math. Soc. Simon Stevin* **27**, 3, pp. 419–430.

Beer, G., Costantini, C., and Levi, S. (2011). Total boundedness in metrizable spaces, *Houston J. Math.* **37**, 4, pp. 1347–1362.

Beer, G. and Di Maio, G. (2010). Cofinal completeness of the Hausdorff metric topology, *Fund. Math.* **208**, 1, pp. 75–85.

Beer, G. and Di Maio, G. (2012). The bornology of cofinally complete subsets, *Acta Math. Hungar.* **134**, 3, pp. 322–343.

Beer, G., García-Lirola, L. C., and Garrido, M. I. (2020). Stability of Lipschitz-type functions under pointwise product and reciprocation, *Rev. R. Acad. Cienc. Exactas Fís. Nat. Ser. A Mat. RACSAM* **114**, 3, Paper No. 120, p. 16.

Beer, G. and Garrido, M. I. (2014). Bornologies and locally Lipschitz functions, *Bull. Aust. Math. Soc.* **90**, 2, pp. 257–263.

Beer, G. and Garrido, M. I. (2015). Locally Lipschitz functions, cofinal completeness, and UC spaces, *J. Math. Anal. Appl.* **428**, 2, pp. 804–816.

Beer, G. and Garrido, M. I. (2016). On the uniform approximation of Cauchy continuous functions, *Topology Appl.* **208**, pp. 1–9.

Beer, G. and Garrido, M. I. (2020). Real-valued Lipschitz functions and metric properties of functions, *J. Math. Anal. Appl.* **486**, 1, 123839, p. 18.

Beer, G., Garrido, M. I., and Meroño, A. S. (2018). Uniform continuity and a new bornology for a metric space, *Set-Valued Var. Anal.* **26**, 1, pp. 49–65.

Beer, G. and Levi, S. (2009a). Strong uniform continuity, *J. Math. Anal. Appl.* **350**, 2, pp. 568–589.

Beer, G. and Levi, S. (2009b). Total boundedness and bornologies, *Topology Appl.* **156**, 7, pp. 1271–1288.

Beer, G. and Segura, M. (2009). Well-posedness, bornologies, and the structure of metric spaces, *Appl. Gen. Topol.* **10**, 1, pp. 131–157.

Beer, G. A., Himmelberg, C. J., Prikry, K., and Van Vleck, F. S. (1987). The locally finite topology on 2^X, *Proc. Amer. Math. Soc.* **101**, 1, pp. 168–172.

Borsík, J. (1988). Mappings that preserve Cauchy sequences, *Časopis Pěst. Mat.* **113**, 3, pp. 280–285.

Borsík, J. (2000). Mappings preserving Cauchy nets, *Tatra Mt. Math. Publ.* **19**, pp. 63–73.

Bourbaki, N. (1966). *Elements of mathematics. General topology. Part 1* (Hermann, Paris; Addison-Wesley Publishing Co., Reading, Mass.-London-Don Mills, Ont.).

Bouziad, A. and Sukhacheva, E. (2019). Preservation of uniform continuity under pointwise product, *Topology Appl.* **254**, pp. 132–144.

Brandi, P., Ceppitelli, R., and Holá, L. (2008). Boundedly UC spaces and topologies on function spaces, *Set-Valued Anal.* **16**, 4, pp. 357–373.

Burdick, B. S. (2000). On linear cofinal completeness, *Topology Proc.* **25**, pp. 435–455.

Cabello Sánchez, J. (2017). $U(X)$ as a ring for metric spaces X, *Filomat* **31**, 7, pp. 1981–1984.

Chaves, M. A. (1985). Spaces where all continuity is uniform, *Amer. Math. Monthly* **92**, 7, pp. 487–489.

Corson, H. H. (1958). The determination of paracompactness by uniformities, *Amer. J. Math.* **80**, pp. 185–190.

Das, P., Pal, S. K., and Adhikary, N. (2020). On certain versions of straightness, *Topology Appl.* **284**, 107369, p. 15.

Doss, R. (1947). On uniformly continuous functions in metrizable spaces, *Proc. Math. Phys. Soc. Egypt* **3**, pp. 1–6.

Dugundji, J. (1966). *Topology* (Allyn and Bacon, Inc., Boston, Mass.).

García-Máynez, A. and Romaguera, S. (1999). Perfect pre-images of cofinally complete metric spaces, *Comment. Math. Univ. Carolin.* **40**, 2, pp. 335–342.

Garrido, M. I. and Jaramillo, J. A. (2004). Homomorphisms on function lattices, *Monatsh. Math.* **141**, 2, pp. 127–146.

Garrido, M. I. and Jaramillo, J. A. (2008). Lipschitz-type functions on metric spaces, *J. Math. Anal. Appl.* **340**, 1, pp. 282–290.

Garrido, M. I. and Meroño, A. S. (2016). On paracompactness, completeness and boundedness in uniform spaces, *Topology Appl.* **203**, pp. 98–107.

Garrido, M. I. and Meroño, A. S. (2012). Some classes of bounded sets in metric spaces, (Univ. Complut. Madrid, Fac. Mat., Madrid), pp. 179–186.

Garrido, M. I. and Meroño, A. S. (2013). Uniformly metrizable bornologies, *J. Convex Anal.* **20**, 1, pp. 285–299.

Garrido, M. I. and Meroño, A. S. (2014). New types of completeness in metric spaces, *Ann. Acad. Sci. Fenn. Math.* **39**, 2, pp. 733–758.

Gupta, L. and Kundu, S. (2020). Functions that preserve certain classes of sequences and locally Lipschitz functions, *Ann. Acad. Sci. Fenn. Math.* **45**, 2, pp. 699–722.

Gupta, L. and Kundu, S. (2021a). Cofinal completeness vis-á-vis hyperspaces, *Rev. R. Acad. Cienc. Exactas Fís. Nat. Ser. A Mat. RACSAM* **115**, 2, p. 18.

Gupta, L. and Kundu, S. (2021b). Cofinal completion vis-á-vis Cauchy continuity and total boundedness, *Topology Appl.* **290**, 107576, p. 12.

Gupta, L. and Kundu, S. (2022). Cauchy-subregular functions vis-à-vis different types of continuity, *Topology Appl.* **312**, 108088, p. 15.

Gutú, O. and Jaramillo, J. A. (2019). Surjection and inversion for locally Lipschitz maps between Banach spaces, *J. Math. Anal. Appl.* **478**, 2, pp. 578–594.

Hejcman, J. (1959). Boundedness in uniform spaces and topological groups, *Czechoslovak Math. J.* **9 (84)**, pp. 544–563.

Hohti, A. (1981). On uniform paracompactness, *Ann. Acad. Sci. Fenn. Ser. A I Math. Dissertationes*, **36**, pp. 1–46.

Holá, L. (1988). Hausdorff metric convergence of continuous functions, in *General topology and its relations to modern analysis and algebra, VI (Prague, 1986), Res. Exp. Math.*, Vol. 16 (Heldermann, Berlin), pp. 263–271.

Holá, L. (1992). Hausdorff metric on the space of upper semicontinuous multifunctions, *Rocky Mountain J. Math.* **22**, 2, pp. 601–610.

Holá, L. and Neubrunn, T. (1988). On almost uniform convergence and convergence in Hausdorff metric, *Rad. Mat.* **4**, 1, pp. 193–202.

Howes, N. R. (1971). On completeness, *Pacific J. Math.* **38**, pp. 431–440.

Howes, N. R. (1995). *Modern analysis and topology*, Universitext (Springer-Verlag, New York).

Hueber, H. (1981). On uniform continuity and compactness in metric spaces, *Amer. Math. Monthly* **88**, 3, pp. 204–205.

Jain, T. and Kundu, S. (2007). Atsuji completions: Equivalent characterisations, *Topology Appl.* **154**, 1, pp. 28–38.

Jain, T. and Kundu, S. (2008). Atsuji completions vis-à-vis hyperspaces, *Math. Slovaca* **58**, 4, pp. 497–508.

Keremedis, K. (2017). Metric spaces on which continuous functions are "almost" uniformly continuous, *Topology Appl.* **232**, pp. 256–266.

Keremedis, K. (2018a). On metric spaces where continuous real valued functions are uniformly continuous and related notions, *Topology Appl.* **238**, pp. 45–53.

Keremedis, K. (2018b). Second countable UC metric spaces are Lebesgue in ZF, *Topology Proc.* **52**, pp. 73–93.

Kundu, S., Aggarwal, M., and Hazra, S. (2017). Finitely chainable and totally bounded metric spaces: Equivalent characterizations, *Topology Appl.* **216**, pp. 59–73.

Kundu, S. and Jain, T. (2006). Atsuji spaces: Equivalent conditions, *Topology Proc.* **30**, 1, pp. 301–325.

Künzi, H.-P. A. and Romaguera, S. (1999). Quasi-metric spaces, quasi-metric hyperspaces and uniform local compactness, in *Proceedings of the "I Spanish-Italian Congress on General Topology and its Applications" (Spanish) (Gandia, 1997)*, Vol. 30, pp. 133–144.

Leung, D. H. and Tang, W.-K. (2017). Functions that are Lipschitz in the small, *Rev. Mat. Complut.* **30**, 1, pp. 25–34.

Lowen-Colebunders, E. (1989). *Function classes of Cauchy continuous maps* (Marcel Dekker, Inc., New York).

Luukkainen, J. (1979). Rings of functions in Lipschitz topology, *Ann. Acad. Sci. Fenn. Ser. A I Math.* **4**, 1, pp. 119–135.

McShane, E. J. (1934). Extension of range of functions, *Bull. Amer. Math. Soc.* **40**, 12, pp. 837–842.

Michael, E. (1951). Topologies on spaces of subsets, *Trans. Amer. Math. Soc.* **71**, pp. 152–182.

Miculescu, R. (2000/01). Approximation of continuous functions by Lipschitz functions, *Real Anal. Exchange* **26**, 1, pp. 449–452.

Monteiro, A. A. and Peixoto, M. M. (1951). Le nombre de Lebesgue et la continuité uniforme, *Portugaliae Math.* **10**, pp. 105–113.

Mrówka, S. G. (1965). On normal metrics, *Amer. Math. Monthly* **72**, pp. 998–1001.

Nadler, S. B., Jr. and West, T. (1981). A note on Lebesgue spaces, *Topology Proc.* **6**, 2, pp. 363–369.

Nagata, J. (1950). On the uniform topology of bicompactifications, *J. Inst. Polytech. Osaka City Univ. Ser. A. Math.* **1**, pp. 28–38.

Naimpally, S. A. (1966). Graph topology for function spaces, *Trans. Amer. Math. Soc.* **123**, pp. 267–272.

Njåstad, O. (1965). On uniform spaces where all uniformly continuous functions are bounded, *Monatsh. Math.* **69**, pp. 167–176.

O'Farrell, A. G. (2004). When uniformly-continuous implies bounded, *Irish Math. Soc. Bull.* **53**, pp. 53–56.

Rainwater, J. (1959). Spaces whose finest uniformity is metric, *Pacific J. Math.* **9**, pp. 567–570.

Rice, M. D. (1977). A note on uniform paracompactness, *Proc. Amer. Math. Soc.* **62**, 2, pp. 359–362.

Romaguera, S. (1998). On cofinally complete metric spaces, *Questions Answers Gen. Topology* **16**, 2, pp. 165–170.

Romaguera, S. and Antonino, J. A. (1993). On Lebesgue quasi-metrizability, *Boll. Un. Mat. Ital. A (7)* **7**, 1, pp. 59–66.

Sendov, B. (1990). *Hausdorff approximations, Mathematics and its Applications (East European Series)*, Vol. 50 (Kluwer Academic Publishers Group, Dordrecht), translated and revised from the Russian.

Smith, J. (1978). Review of "A note on uniform paracompactness" by Michael D. Rice, *Math. Rev.* **55**, #9036.

Snipes, R. F. (1977). Functions that preserve Cauchy sequences, *Nieuw Arch. Wisk. (3)* **25**, 3, pp. 409–422.

Snipes, R. F. (1981). Cauchy-regular functions, *J. Math. Anal. Appl.* **79**, 1, pp. 18–25.

Toader, G. (1978). On a problem of Nagata, *Mathematica (Cluj)* **20(43)**, 1, pp. 77–79.

Waterhouse, W. C. (1965). On *UC* spaces, *Amer. Math. Monthly* **72**, pp. 634–635.

Williamson, R. and Janos, L. (1987). Constructing metrics with the Heine-Borel property, *Proc. Amer. Math. Soc.* **100**, 3, pp. 567–573.

Wong, Y. M. (1972). The Lebesgue covering property and uniform continuity, *Bull. London Math. Soc.* **4**, pp. 184–186.

Index

$AC(X)$, 25
$CV(X,Y)$, 35
$FV(X,Y)$, 58
$LLV(X,Y)$, 86
ε-chain, 4
ν-bounded, 29
t-bounded, 57

almost bounded function, 35
almost nowhere locally compact, 24
almost uniformly continuous function, 36
almost uniformly continuous space, 36
Atsuji space, 3
AUC space, 36

BC-regular, 8
bornology, 64
Bourbaki-asymptotic sequences, 102
Bourbaki-Cauchy sequence, 5
Bourbaki-complete space, 5

Cauchy equivalent, 52
Cauchy filter, 19
Cauchy-Lipschitz, 11
Cauchy-regular function, 8
Cauchy-subregular, 14
CBC-regular, 8
CC-regular, 8
chainable, 4
cofinally Čech complete, 22
cofinally asymptotic sequences, 102
cofinally Bourbaki-asymptotic sequences, 102
cofinally Bourbaki-Cauchy sequence, 5
cofinally Bourbaki-complete space, 5
cofinally Cauchy sequence, 1
cofinally complete space, 2
cofinally small, 47

discrete set, 4

finitely chainable, 4

gap, 23

Hausdorff metric topology, 75

isolation functional, 3

Kuratowski measure of non-compactness, 24

Lipschitz, 11
Lipschitz in the small, 11
local compactness functional, 24
local finiteness functional, 47
local total boundedness functional, 50
locally compact, 6
locally finite topology, 76
locally Lipschitz, 12
locally totally bounded, 50

PC-regular, 8
preparacompact, 21
proximal topology, 76
pseudo-Bourbaki-Cauchy sequence, 121
pseudo-Cauchy sequence, 3

regularly bounded, 21

strongly cofinally complete, 48
strongly uniformly continuous, 14

topology of uniform convergence, 77

UC set, 110
UC space, 3
uniform paracompactness, 21
uniformly asymptotic sequences, 102
uniformly discrete set, 4
uniformly locally bounded, 28
uniformly locally compact, 6
uniformly locally Lipschitz, 11
uniformly locally totally bounded, 50

Vietoris topology, 76

weakly Cauchy filter, 19

www.ingramcontent.com/pod-product-compliance
Lightning Source LLC
Chambersburg PA
CBHW050643190326
41458CB00008B/2394
* 9 7 8 9 8 1 1 2 7 2 6 5 3 *